❖ 49 ❖ Easy Electronic Projects for the 747 Dual Op Amp

❖ 49 ❖ Easy Electronic Projects for the 747 Dual Op Amp

Delton T. Horn

TAB BOOKS
Blue Ridge Summit, PA

FIRST EDITION
FIRST PRINTING

© 1991 by **TAB BOOKS**
TAB BOOKS is a division of McGraw-Hill, Inc.

Printed in the United States of America. All rights reserved. The publisher takes no responsibility for the use of any of the materials or methods described in this book, nor for the products thereof.

Library of Congress Cataloging-in-Publication Data

Horn, Delton T.
    49 easy electronic projects for the 747 dual op amp / by Delton T. Horn.
      p.   cm.
    Includes index.
    ISBN 0-8306-7458-6 (hard)   ISBN 0-8306-3458-4 (pbk.)
    1. Operational amplifiers—Amateurs' manuals. I. Title.
II. Title: Forty nine easy electronic projects for the 747 dual op amp.
TK9966.H662 1990
621.39'5—dc20   90-11231
                                              CIP

TAB BOOKS offers software for sale. For information and a catalog, please contact TAB Software Department, Blue Ridge Summit, PA 17294-0850.

Questions regarding the content of this book should be addressed to:

Reader Inquiry Branch
TAB BOOKS
Blue Ridge Summit, PA 17294-0850

Acquisitons Editor: Roland S. Phelps
Technical Editor: B. J. Peterson
Production: Katherine G. Brown
Book Design: Jaclyn J. Boone

Contents

List of projects *vii*

Introduction *ix*

❖ **1 The basic operational amplifier and the 747** *1*
How an op amp works *2*
The 741 op amp *10*
The 747 dual op amp *14*

❖ **2 Operational circuits** *17*
Summing amplifier *17*
Absolute value circuit *20*
Window comparator *24*
Magnitude detector *27*
Peak detector *31*
Inverting peak detector *34*
Peak detector with manual reset *35*
Logarithms *37*
Multiplication circuit *40*
Divider circuit *41*
Exponent circuit *44*

❖ **3 Audio projects** *49*
Audio preamp *49*
Stereo magnetic cartridge preamp *51*
Tape head preamp *53*
Audio tone control *54*
Audio mixer *55*

❖ **4 Signal generator projects** *59*
Quadrature oscillator *60*
Triangle wave generator *63*
Two-frequency oscillator *66*
Sawtooth wave generator *69*

❖ 5 Filter projects — 75
Third-order low-pass filter *77*
Fourth-order high-pass filter *82*
Bandpass filter *85*
60 Hz notch filter *90*
State variable filter *92*
Speech filter *96*

❖ 6 Test equipment — 99
Differential voltmeter *99*
Null voltmeter *102*
LED null indicator *103*
Decibel meter *104*
Rms to dc converter *105*

❖ 7 Modulation projects — 109
FM signal generator *110*
Pulse amplitude modulator *113*
Pulse width modulator *117*

❖ 8 Pulse circuits — 119
One-shot timer *123*
Dc coupled bistable multivibrator *125*
Ac coupled bistable multivibrator *126*
Astable multivibrator *128*
Comparing digital and analog circuitry *130*
Simple D/A converter *132*
R–2R D/A converter *135*
Improved D/A converter *137*

❖ 9 Miscellaneous projects — 141
Dual-polarity voltage regulator *141*
Precision full-wave rectifier *145*
High input impedance inverting amplifier *149*
Amplifier with differential inputs and outputs *152*
Deadspace circuit *154*
Series limiter *157*
Low-impedance instrumentation amplifier *161*
High-impedance instrumentation amplifier *163*
Adjustable gain instrumentation amplifier *166*

Index — *171*

Projects

Project	Description	Page
1	Summing amplifier	17
2	Absolute value circuit	20
3	Window comparator	24
4	Magnitude detector	27
5	Peak detector	31
6	Inverting peak detector	34
7	Peak detector with manual reset	35
8	Multiplier	37
9	Divider	41
10	Exponent circuit	44
11	Audio preamp	49
12	Magnetic cartridge stereo preamp	51
13	Tape head preamp	53
14	Audio tone control	54
15	Mixer	55
16	Quadrature oscillator	60
17	Triangle wave generator	63
18	Two-frequency oscillator	66
19	Sawtooth wave generator	69
20	Third-order low-pass filter	77
21	Fourth-order high-pass filter	82
22	Bandpass filter	85
23	60 Hz notch filter	90
24	State variable filter	92
25	Speech filter	96

List of projects

26	Differential voltmeter	*99*
27	Null voltmeter	*102*
28	LED null indicator	*103*
29	Decibel meter	*104*
30	Rms-to-dc converter	*105*
31	FM signal generator	*110*
32	Pulse amplitude modulator	*113*
33	Pulse width modulator	*117*
34	One-shot timer	*123*
35	Dc coupled bistable multivibrator	*125*
36	Ac coupled bistable multivibrator	*126*
37	Astable multivibrator	*128*
38	Simple D/A converter	*132*
39	R–2R D/A converter	*135*
40	Improved D/A converter	*137*
41	Dual-polarity voltage regulator	*141*
42	Precision full-wave rectifier	*145*
43	High input impedance inverting amplifier	*149*
44	Amplifier with differential inputs and outputs	*152*
45	Deadspace circuit	*154*
46	Series limiter	*157*
47	Low-impedance instrumentation amplifier	*161*
48	High-impedance instrumentation amplifier	*163*
49	Adjustable gain instrumentation amplifier	*166*

Introduction

OPERATIONAL AMPLIFIERS, OR *OP AMPS*, ARE PROBABLY THE MOST popular kind of ICs around today. They can be used in countless applications, including analog computation, audio amplification, signal generation, modulation, and comparators.

If one op amp is so powerful and versatile, think of how much more powerful and useful two op amps can be. More sophisticated applications are possible when multiple op amps are employed in a circuit.

The most popular op amp IC on the market is probably the 741. This device has become the standard by which other op amp chips are compared. More recent devices may have better specifications, but the improvements are usually spoken of in comparison to the 741. The 741 is a handy work horse op amp. It is inexpensive, widely available, and easy to use.

In this book we will be concentrating on the 747 dual op amp IC. The 747 contains two 741 op amp circuits.

In multiple op amp projects, like those described in this book, separate op amp chips (such as the 741) can be used, but this is inefficient, unnecessarily increasing the part count and size of the project. A cleaner solution is to use a dual op amp device, such as the 747.

Another important advantage (in some circuits) for using a dual op amp device is that two op amp circuits etched on a single chip of silicon will be very closely matched in all of their characteristics. Such matching is highly desirable in many critical applications.

Although a number of books have been published on operational amplifiers and the 741, I believe this one is unique in its focus on the 747 dual op amp IC. This book features projects for the 747 dual op amp. Suggestions are given on modifying and customizing many of the projects. You certainly don't have to be an electronics expert to build and experiment with these projects. If you know how to solder and how to read a schematic diagram, you should have no trouble building any of these 49 multiple op amp projects.

❖ 1
The basic operational amplifier and the 747

MOST MODERN ELECTRONIC HOBBYISTS ARE QUITE FAMILIAR WITH the standard operational amplifier, or op amp. It is probably the most widely used and versatile of all integrated circuit types around today. The op amp has countless applications in such diverse areas as amplification, filtering, mathematical functions, audio applications, test equipment, signal generation, and many others. Many different op amp ICs are available to the electronics hobbyist, from the very basic and inexpensive, to some very complex units.

Operational amplifier circuits originally were designed primarily to perform mathematical operations in analog computation circuits. This intended function certainly explains the name.

Early op amp circuits that used discrete components were complex, bulky, and expensive, so they rarely were used unless absolutely essential. With the development of the integrated circuit op amp in the late sixties and early seventies, this situation changed drastically. As costs and sizes dropped to almost negligible levels, countless new applications were found for the op amp. Today the operational amplifier IC is considered an all-purpose device.

Sooner or later, specialized versions of all standard components start to show up, and the op amp is certainly no exception to this rule. One particularly handy variation on the basic operational amplifier is the multiple op amp, with two or more electrically separate op amp units in a single IC package.

2 The basic operational amplifier and the 747

The primary focus of this book will be the 747 dual op amp chip. Each 747 contains two complete op amp circuits. Each of the 747's op amp stages is electrically and functionally identical to the popular 741 op amp.

HOW AN OP AMP WORKS

On the most basic level, an operational amplifier is a dc voltage differential amplifier with extremely high gain. It also can be operated with ac signals over a very wide frequency range [typically from dc (zero Hertz) to several Megahertz (1 MHz = 1,000,000 Hz)]. In a practical circuit, the operational amplifier's frequency response and gain can be controlled with external feedback networks.

The conventional schematic symbol for an op amp is shown in Fig. 1-1. Notice that this device has two separate voltage inputs, V_1 (noninverting) and V_2 (inverting). The noninverting input is indicated by a plus sign (+) on op amp schematic diagrams, and the inverting input by a minus sign (−). The op amp's output signal is a voltage equal to:

$$V_{out} = G \times (V_1 - V_2)$$

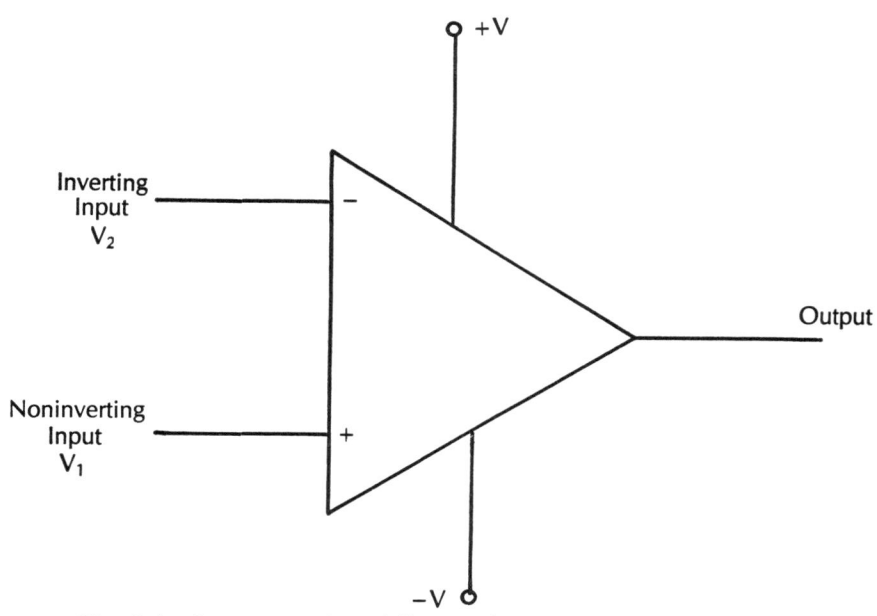

Fig. 1-1 *An op amp is a differential amplifier for dc voltages.*

where *G* is the gain of the device.

The open loop (no feedback) gain of an op amp is normally extremely high. In most practical circuits, positive feedback is used to reduce the gain to a significantly lower level. This is done by setting up a closed feedback path between the op amp's output and its inverting input.

In use, the differential inputs of an operational amplifier can be used in a wide variety of ways. For convenience in the following discussion, assume that both of the input signals are positive. There are three possible combinations:

- The inverting input voltage (V_2) is greater than the noninverting input voltage (V_1).
- The inverting input voltage (V_2) is less than the noninverting input voltage (V_1).
- The inverting input voltage (V_2) is equal to the noninverting input voltage (V_1).

In the first case ($V_2 > V_1$), the output voltage will be negative (inverted). With the second combination ($V_2 < V_1$) the output voltage will be positive (noninverted). Finally, in the third possible combination ($V_2 = V_1$) the two input voltages will cancel each other out and the output voltage will be zero.

As an example of how the op amp is used, one of the most basic op amp circuits, the *inverting amplifier*, is illustrated in Fig. 1-2. Notice that only one signal input is shown for this circuit.

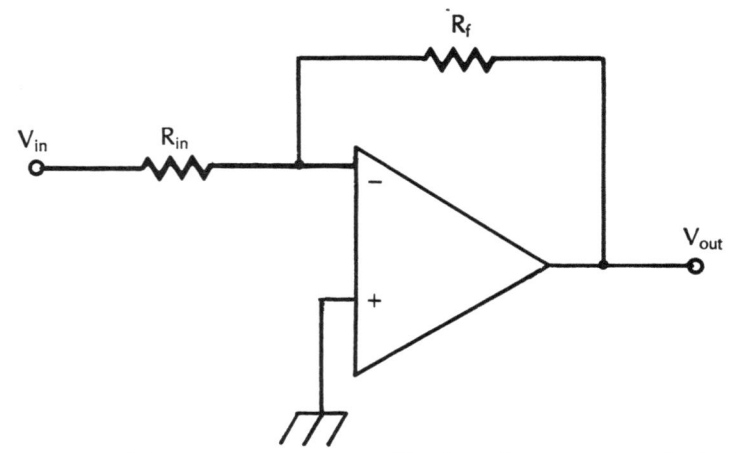

Fig. 1-2 *One of the most common applications for an op amp is the inverting amplifier.*

4 The basic operational amplifier and the 747

The operational amplifier's noninverting input is simply grounded in this circuit. This is exactly the same as feeding a constant signal of 0 volts to this input.

Since subtracting zero from any value results in the original value, we can simply ignore the second input in this particular application. The output voltage is controlled only by the signal at the inverting input and the circuit gain:

$$\begin{aligned} V_{out} &= G \times (V_1 - V_2) \\ &= G \times (0 - V_2) \\ &= G \times -V_2 \end{aligned}$$

The negative sign indicates that the output signal is 180 degrees out of phase with the input signal. This is the effect of using the inverting input. By definition, the inverting input of an op amp inverts (or reverses) the signal polarity. If the input signal is positive, then the output signal will be negative, and vice versa.

There are two resistors in a basic inverting amplifier circuit. One (R_{in}) is the input resistor, and the other (R_f) is the feedback resistor. The relative values of these two resistors determine the closed loop gain of the op amp in the circuit. In some cases, a third resistor will be added between the op amp's noninverting input and ground. This extra resistor improves the stability of the circuit. The gain formula for the basic inverting amplifier circuit is quite simple:

$$\begin{aligned} G &= -R_f/R_{in} \\ V_{out} &= -V_{in} R_f/R_{in} \end{aligned}$$

where: R_f is the value of the feedback resistor
R_{in} is the value of the input resistor
V_{in} is the input voltage
V_{out} is the output voltage
G is the amplifier gain

The negative sign in this equation simply indicates the polarity inversion that occurs in this circuit. The output voltage always has the opposite polarity as the input voltage.

As a typical example, let the input resistor (R_{in}) have a value of 10K (10,000 ohms), and the feedback resistor (R_f) have a value of 100K (100,000 ohms). In this case, the circuit's gain works out to:

$$G = -R_f/R_{in}$$

$$= -100{,}000/10{,}000$$
$$= -10$$

As long as the feedback resistor has a larger value than the input resistor, the gain will be positive (greater than unity), and the circuit will amplify the input signal. The output voltage will be larger (with a reversed polarity) than the input voltage. On the other hand, if the feedback resistor is given a smaller value than the input resistor, the gain will be negative (less than unity), and the circuit will attenuate the input signal. In this case the output voltage will be smaller (with a reversed polarity) than the input voltage.

There is one other possibility. If the input resistor and the feedback resistor have identical values, the gain will be unity (1). The signal will be neither amplified nor attenuated. The output voltage will be equal to the input voltage, except for the inversion of the polarity. This is true even if the value given to both resistors is zero. In other words, the resistors may be omitted from the circuit altogether, as shown in Fig. 1-3. This circuit is known as an *inverting voltage follower*. The gain is − 1.

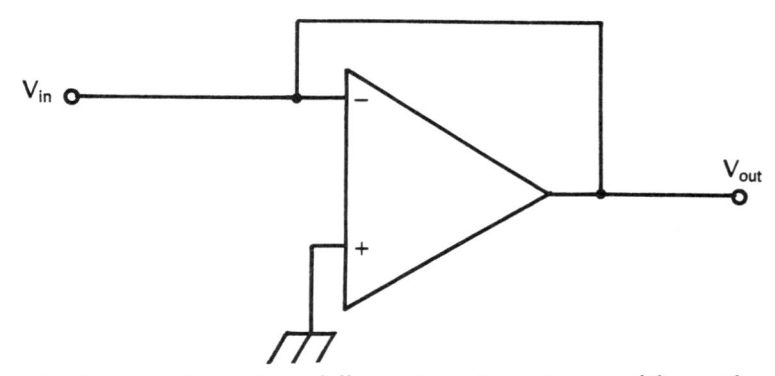

Fig. 1-3 *An inverting voltage follower is an inverting amplifier with a gain of − 1.*

You might wonder what would happen if one of the two resistors was used, but the other was omitted (effective resistance of zero ohms). Either of the two possible combinations results in a *disallowed condition*. This does not mean there is any law against it. You certainly won't be punished for trying such a circuit. But the resulting circuit will not be functional. To see why, try an example for each of the two possible disallowed conditions. In each case, assume that the single resistor has a value of 10K (10,000 ohms). As you'll soon see, the actual resistance used doesn't matter at all.

First consider what happens if a feedback resistor is used, but the input resistor is omitted (0 ohms). The gain in this case works out to:

$$G = -R_f/R_{in}$$
$$= -10,000/0$$
$$= -\infty$$

Any finite value divided by zero always results in infinity (∞).

Actually, in a practical circuit of this type, the gain really won't be infinite. That would be impossible. The circuit will exhibit the maximum open loop gain of the operational amplifier used.

Now consider the other disallowed condition. This time, have an input resistor, but no feedback resistor (0 ohms). In this case, the gain works out to:

$$G = -R_f/R_{in}$$
$$= -0/10,000$$
$$= 0$$

Remember that the output voltage is equal to the input voltage, multiplied by the circuit gain:

$$V_{out} = V_{in} \times G$$
$$= V_{in} \times 0$$
$$= 0$$

The output voltage always will be zero, regardless of the value of the input voltage. Obviously, such a circuit could serve no useful purpose.

Another basic op amp circuit is the *noninverting amplifier* circuit, illustrated in Fig. 1-4. In concept, this circuit is similar to the inverting amplifier shown in Fig. 1-2. Note that in Fig. 1-4 the op amp schematic is inverted from the one in Fig. 1-2. The schematic can be drawn in either way, you must pay attention to the labeling. Again, just one of the operational amplifier's differential inputs is used, the noninverting input in this case. This circuit also employs negative feedback by returning some of the output signal to the inverting input to reduce the open loop gain.

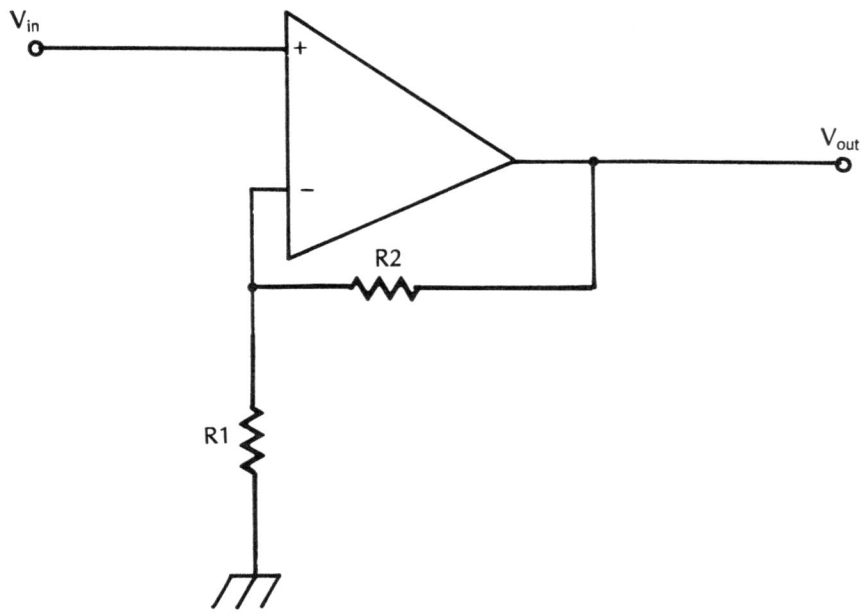

Fig. 1-4 *Another basic op amp circuit is the noninverting amplifier.*

Notice that the inverting input must be used for the negative feedback. The polarity inversion (phase shift) of the inverting input is an essential element of negative feedback. Using the noninverting input for the feedback will cause the feedback signal to add to the original input signal and increase, rather than decrease, the open loop gain. At best, the result will be circuit instability and/or uncontrolled oscillations. In many cases, the op amp chip and possibly some other components in the circuit could be damaged or destroyed.

The gain equation for a noninverting amplifier circuit is somewhat more complex than the one for an inverting amplifier. Resistors R1 and R2 form a voltage divider that determines how much voltage will be fed back. The noninverting gain equation is:

$$G = (1 + R_2/R_1)$$

Notice that there is no negative sign in this equation because the noninverting input does not reverse the signal polarity. The output signal is in phase with the input signal. Notice also that the gain of a noninverting amplifier circuit must always be at least equal to unity. Negative gains (attenuation) are not possible with this circuit.

As an example, let the resistor R_1 have a value of 10K (10,000 ohms), and the resistor R_2 have a value of 22K (22,000 ohms). In this case, the circuit's gain works out to:

$$G = 1 + (22,000/10,000)$$
$$= 1 + 2.2$$
$$= 3.2$$

The lowest possible gain for a noninverting amplifier circuit is achieved when full V_{out} voltage is fed back, as shown in Fig. 1-5. This effectively reduces the R_2 resistance value to zero. In this case, output voltage works out to:

$$V_{out} = (1 + 0) \times V_{in}$$
$$= 1 \times V_{in}$$
$$= V_{in}$$

Under these circumstances, we have unity gain. The output signal is identical to the input signal. Such a circuit is used as a buffer or an impedance matcher. It is called a *noninverting voltage follower*.

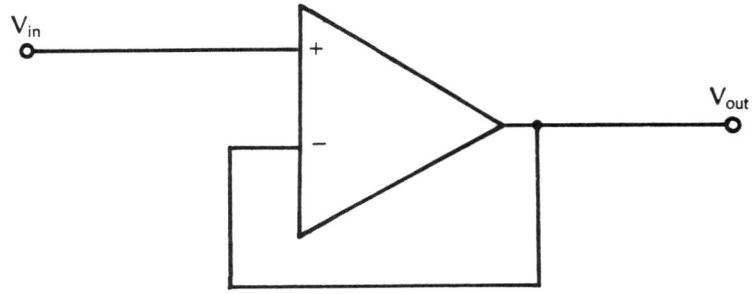

Fig. 1-5 *The noninverting voltage follower is an amplifier with unity gain.*

Notice that unity gain is achieved in a noninverting amplifier circuit when the R_2 resistance is zero. If equal nonzero resistances are used, the gain will always work out to be 2. For example, assume both the resistors have values of 10K (10,000 ohms). In this case, the circuit's gain works out to:

$$G = 1 + (10,000/10,000)$$
$$= 1 + 1$$
$$= 2$$

This is true for any nonzero resistance value.

If both of the operational amplifier's inputs are used, we have a *differential amplifier* circuit, as illustrated in Fig. 1-6. Once again, negative feedback is used to reduce the open loop gain of the operational amplifier.

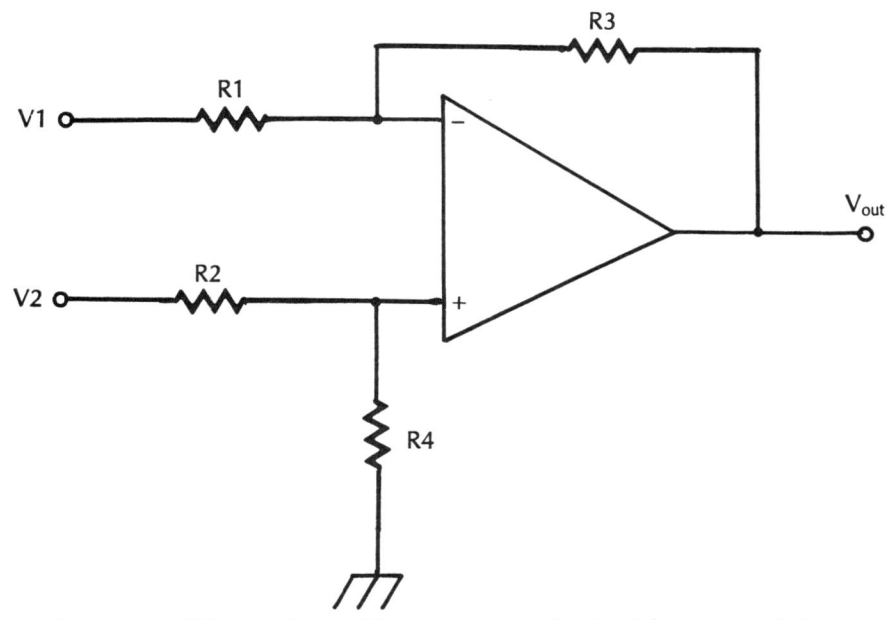

Fig. 1-6 A differential amplifier circuit uses both of the op amp's inputs.

In most cases, resistance R_3 will be equal to resistance R_4, and resistance R_1 will be equal to resistance R_2. These equalities don't always have to be true, but they do significantly simplify the circuit design in many practical applications. There is rarely (if ever) any good reason for the circuit designer not to follow these resistance unities.

In any case, for a true differential amplifier, the $R_3:R_1$ and $R_4:R_2$ ratios must be equal. That is:

$$R_3/R_1 = R_4/R_2$$

The circuit will still function even if these ratio equalities are not maintained, but the signals at the inverting and the noninverting inputs will be subjected to differing amounts of gain, which would be undesirable, or inconvenient in most practical applications.

10   The basic operational amplifier and the 747

Generally speaking, it is easiest on the circuit designer to maintain the resistance value equalities just mentioned:

$$R_1 = R_2$$
$$R_3 = R_4$$

These resistance ratios determine the gain of the amplifier:

$$G = R_3/R_1 = R_4/R_2$$

Assuming the resistance ratio equalities are maintained, the output voltage will be equal to the difference between the two input voltages, multiplied by the gain. That is:

$$V_{out} = G \times (V_1 - V_2)$$

which is the general op amp gain equation given earlier in this section.

Of course a great many other operational amplifier circuits are possible, but most are, in one way or another, variants on the basic circuits presented here. In many op amp applications, either the input resistor or the feedback resistor (or sometimes both) will be replaced with some other type of component, such as a capacitor, a diode, or a transistor. This will have a profound effect on the operation of the circuit.

Notice that in working with an op amp circuit, you usually are concerned only with signals in the form of voltages. Most op amp ICs, including the 741 and the 747, require a dual power supply. That is, two supply voltages are needed—one positive with respect to ground, and the other negative with respect to ground. Usually the dual supply voltages are symmetrical with respect to ground. That is, the positive supply voltage is equal to the negative supply voltage, except for the reversed polarity.

The ideal op amp has infinite input impedance, and infinite open loop gain. The output impedance of an ideal operational amplifier is zero. Of course, the ideal op amp really doesn't exist. Practical operational amplifier devices exhibit very high (but finite) input impedances and open loop gains, and very low (but not quite zero) output impedances.

THE 741 OP AMP

The de facto standard op amp is the 741. This popular device is probably the most widely used IC around. Most other op amp devices are designed to be compatible with the 741. Many op amp specifications commonly are compared with those of the 741.

The 741 operational amplifier is available in a number of standard package styles. Most modern op amp ICs are designed to be pin-for-pin compatible with the 741. That is, each pin performs the same function as on the 741.

The most common packaging style is the 8-pin dual inline package (DIP), shown in Fig. 1-7. Notice that pin 8 is not used. It is not electrically connected to anything within the chip.

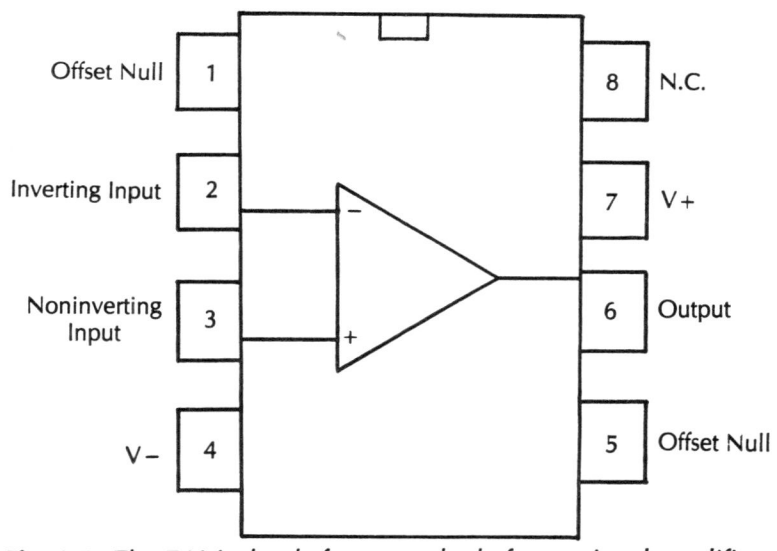

Fig. 1-7 *The 741 is the de facto standard of operational amplifiers.*

Pins 4 and 7 are the IC's power supply connections. The 741 requires a dual polarity power supply. The positive supply voltage is connected to pin 7, and the negative supply voltage is fed to pin 4. The common (ground) point is not connected directly to the chip, although it usually is used as part of the external circuitry.

Pin 2 is the inverting input and pin 3 is the noninverting input. The output of the operational amplifier is pin 6.

The remaining two pins on this chip (pins 1 and 5) are left unconnected in most circuits. These pins are used for offset null connections. When the differential voltage between the inverting input and the noninverting input is zero, the output should be

12 The basic operational amplifier and the 747

exactly zero. Minor errors within the op amp's internal circuitry might result in some small offset from zero under this condition. The offset null connections are used to compensate for such offset errors in high-precision applications.

The 741 operational amplifier IC is also available in an 8-pin round can. The pin-out diagram for this device is shown in Fig. 1-8.

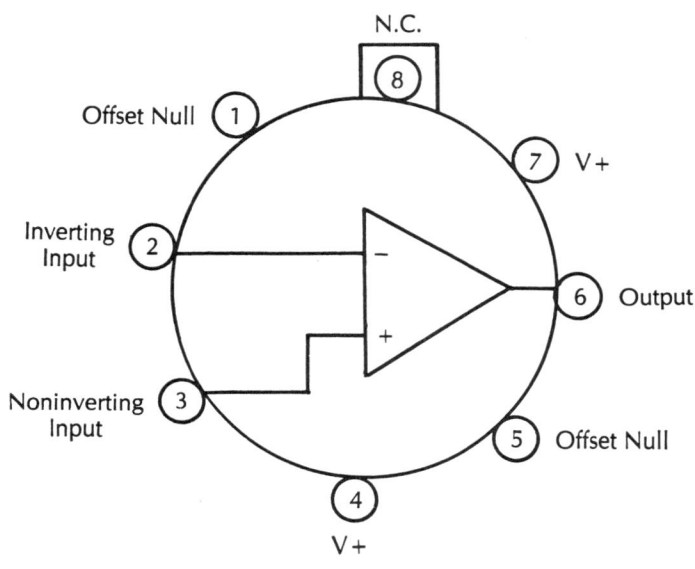

Fig. 1-8 *The 741 op amp in an 8-pin round can package.*

The 14-pin DIP housing illustrated in Fig. 1-9 isn't too widely used, but you may run across it occasionally. The 6 extra pins are not used. This makes a total of seven out of the 14 pins on this device that do not make any electrical connections within the chip. In some applications, the extra pins may be used for heat-sinking purposes.

Unlike some earlier op amp devices, the 741 is internally frequency compensated. An uncompensated op amp may tend to become unstable at high frequencies, unless an external stabilizing capacitor is used. An external frequency compensation capacitor is not needed with the 741.

Both a positive supply voltage and a negative supply voltage are necessary for the 741 to function. Both supply voltages should normally be equal (with respect to ground), except for the

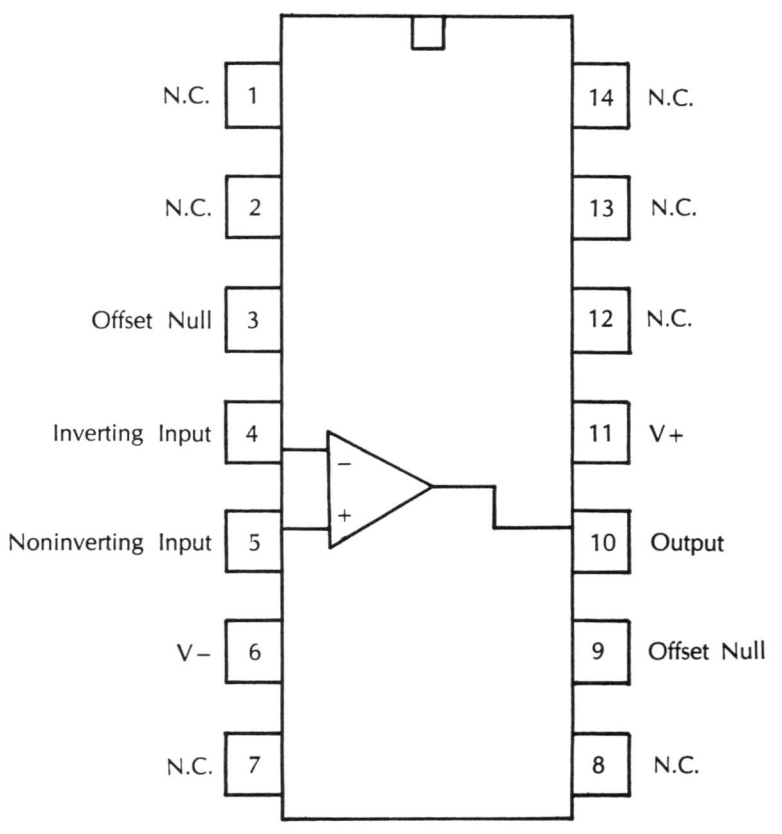

```
                    ┌───┐
         N.C.  │ 1        14 │ N.C.

         N.C.  │ 2        13 │ N.C.

    Offset Null │ 3        12 │ N.C.

 Inverting Input │ 4        11 │ V+

Noninverting Input │ 5       10 │ Output

          V−  │ 6         9 │ Offset Null

         N.C.  │ 7         8 │ N.C.
```

Fig. 1-9 *The 741 op amp in a 14-pin DIP housing.*

opposite polarities. This is sometimes stated as "a dual polarity power source, symmetrical around true ground potential (zero volts)."

The power supply for a 741 op amp may be anything up to ±30 volts. The most popular (and reliable) supply voltages for this device are ±12 volts and ±15 volts. Circuit operation may be unreliable if the supply voltages are less than ±9 volts.

The open-loop (no feedback) gain of the 741 is typically rated as 200,000. That is, if the differential input voltage seen by the inputs is 1 millivolt (0.001 volt), the output voltage will (theoretically) be equal to 200 volts. Actually, the output voltage will be clipped to a value just slightly below the supply voltage. An op amp's output voltage can never exceed its supply voltage. Lesser gains can be obtained by putting an external resistance in

14 The basic operational amplifier and the 747

the negative feedback loop, between the op amp's output and the inverting input.

Due to unavoidable variations during manufacture, there will be some fluctuation in the actual open loop gain (and certain other specifications) from device to device. The open loop gain of a 741 is typically 200,000, but it may be as low as 50,000 (but no lower).

The input impedance of the 741 has a typical rating of 6 megohms (6 million ohms). On some devices, the actual input impedance may be as low as 1 megohm (1 million ohms). The typical output impedance rating for the 741 is 75 ohms.

Another significant specification for an operational amplifier is the *Common Mode Rejection Ratio* (CMRR). If the exact same signal is simultaneously fed to both the inverting input and the noninverting input, the resulting differential voltage should be zero. Because of minor inaccuracies in the op amp's internal circuitry, some of the common mode signal may get through. The CMRR is a measurement of how well common mode signals are blocked by the op amp. The 741 is rated for a minimum CMRR rating of 70 dB. The typical CMRR value for this device is 90 dB.

One final important specification for op amps is the *slew rate*. This is a measurement of how rapidly the output voltage can respond to quick changes in the input voltages. The slew rate is usually given as volts per microsecond. The typical slew rate specification for the 741 is 0.5 V/μS. That is, the output is guaranteed to be able to accurately follow voltage changes as fast as 1/2 volt in a microsecond. (1 microsecond = 0.000001 second.) The slew rate sets the maximum reliable operating frequency for an op amp. A 741 can usually be operated at frequencies up to 1 MHz (1,000,000 Hz) without problems or instability.

THE 747 DUAL OP AMP

The 747, which is the main subject of this book, is a fourteen pin DIP device, containing two 741 type op amp circuits on a single chip. The pin-out diagram for this device is shown in Fig. 1-10.

Notice that the negative supply voltage connection (pin 4) is the only pin on the chip used in common by both internal op amps. There are even separate connections for the positive supply voltage—pin 13 for op amp A, and pin 9 for op amp B. In

The 747 dual op amp

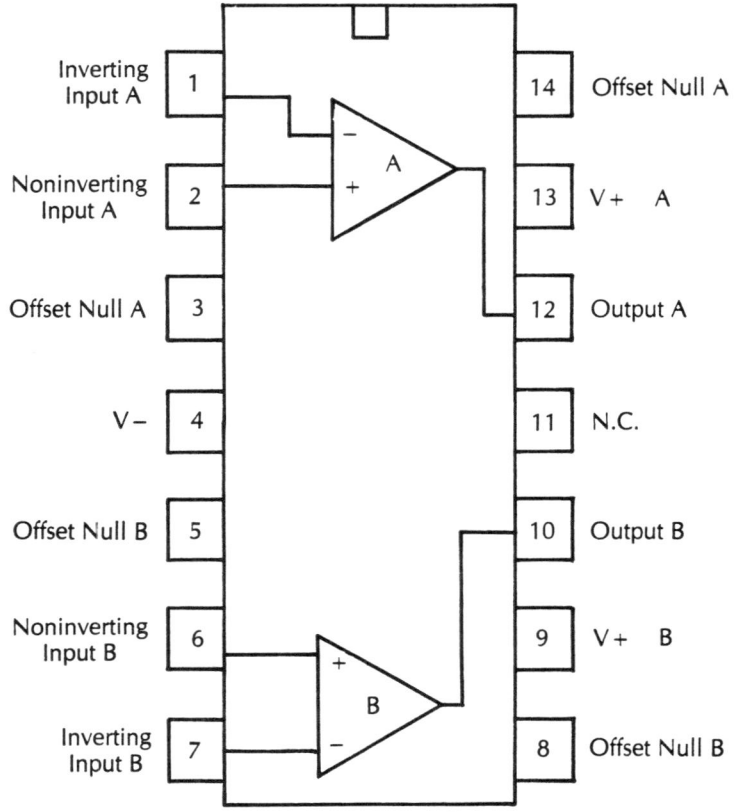

Fig. 1-10 *The 747 dual op amp IC includes two 741 type operational amplifiers in a single 14-pin DIP package.*

most practical applications, however, these two pins are shorted together, and a single positive supply voltage is used for both op amps. This is true for all the projects presented in this book.

To keep the circuit diagrams for the projects as simple as possible, the power supply connections will not be shown in any of the project schematics. **Remember, the negative supply voltage must be connected to pin 4, and the positive supply voltage must be connected to pins 9 and 13, or the circuit will not work.** In some cases, the chip could be damaged. The power supply connections are always assumed. In technical literature, an op amp's power supply connections are sometimes explicitly shown, and sometimes they are simply assumed. Just remember, they are always necessary, whether they are actually shown in the schematic diagram or not.

The electrical specifications for each of the 747's internal op amps are identical to those of the 741 single op amp IC. Two 741s can be directly substituted for a 747, or vice versa, without any other changes in the circuitry.

Whenever two or more circuits are etched onto a single silicon chip, there will inevitably be some leakage between them. The *channel separation* figure given in the specification sheet for the 747 is a measurement of the degree of isolation between the two on-chip op amps. If a signal is applied to op amp A, some of the same signal will also pass through op amp B, but its amplitude will be significantly reduced. The rated channel separation value for the 747 is 120 dB. The leakage signal in op amp B will be 120 dB lower than the original signal fed to op amp A. Unless the input signal is very large, this leakage signal will be small enough that it can reasonably be ignored in the vast majority of applications. In any application where signal leakage between op amp stages might be a significant problem, separate op amp devices should be used.

The 741/747 op amps give quite good performance for most applications; however, they do tend to be a little noisy in some applications. That is, the op amp's internal circuitry will generate a certain amount of random noise, or hiss. If this is objectionable, substitute a high-grade, low-noise op amp for the 747s or 741s.

Compared to some more recent devices, the 747 and the 741 are also pretty limited in terms of frequency response. Again, in critical applications, it may be desirable to substitute an op amp device with better ratings. However, you are not likely to run into any frequency response problems with any of the projects described in this book.

❖2
Operational circuits

OPERATIONAL AMPLIFIERS, OR OP AMPS, WERE ORIGINALLY INVENTED to be used as dedicated (nonprogrammable) analog computers. That is, they were intended to perform various mathematical operations.

In its most basic form, an op amp is a differential amplifier. It calculates the difference between two input values (voltages). In other words, it performs the mathematical operation of subtraction. The value (voltage) at the inverting input is subtracted from the value (voltage) at the noninverting input.

In this chapter, we will put the op amp to work performing other mathematical operations—some of them fairly sophisticated.

SUMMING AMPLIFIER

One of the most basic of all mathematical operations is addition, and an op amp can perform this function quite easily. The circuit shown in Fig. 2-1 adds together the values (or voltages) applied to the three inputs. (This circuit can easily be expanded for more than three inputs.) A typical parts list for this project is given in Table 2-1.

The first stage of this circuit (IC1A) is an inverting summing amplifier. The output voltage is equal to the inverted sum of all the input voltages, multiplied by the amplifier gain. For use as a true adding circuit, unity gain should be used. In addition, for most applications, the inputs should not be weighted (or subjected to differing gains). In other words, the three input resistors (R1 through R3) should all have equal values. To provide unity gain, the feedback resistor (R4) should have the same value as

18 Operational circuits

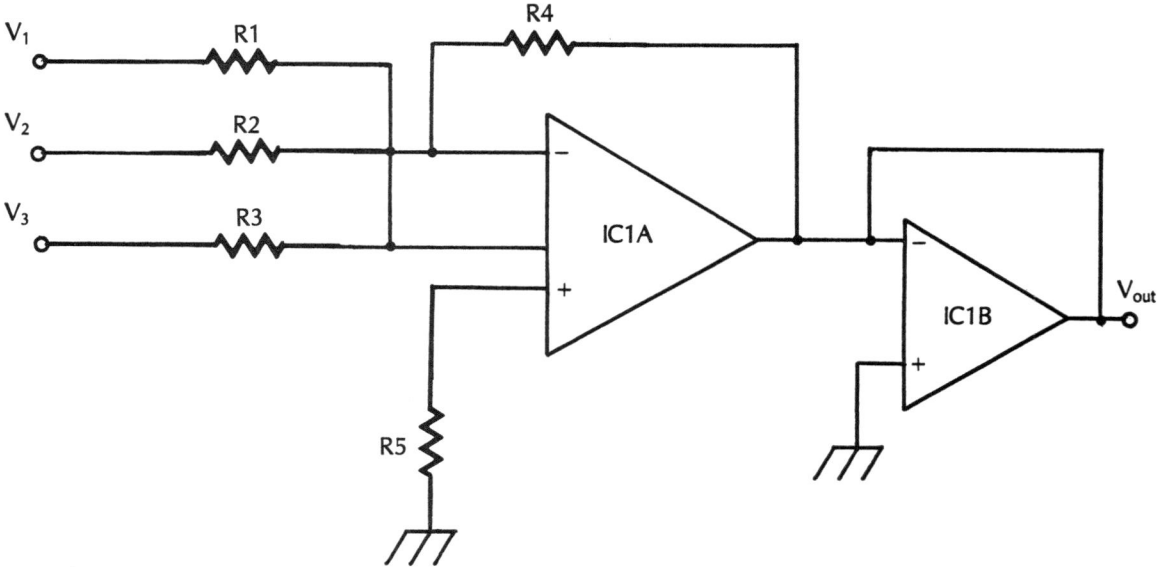

Fig. 2-1 Project 1: summing amplifier.

Table 2-1. Parts List for Project 1: Summing Amplifier.

Part	Description
IC1	747 dual op amp
R1 – R4	10K 5% 1/4-watt resistor
R5	3.3K 5% 1/4-watt resistor

each of the input resistors. That is:

$$R_1 = R_2 = R_3 = R_4$$

The second stage (IC1B) is just an inverting voltage follower. This stage reinverts the signal to restore the original input polarity. If all input voltages are positive, the output of this circuit will also be positive.

There is just one other component to select in this circuit: resistor R5. This resistor compensates for any possible input bias errors, and should have a value approximately equal to the parallel combination of all of the input resistors:

$$R_5 = \frac{1}{1/R_1 + 1/R_2 + 1/R_3}$$

Since the input resistors all have equal values ($R_1 = R_2 = R_3$), I can simplify this equation to the following form:

$$R_5 = \frac{1}{3/R_1}$$
$$= \frac{R_1}{3}$$

Resistor R5 should have a value equal to one-third of one of the input resistors, assuming the circuit has three inputs. Additional inputs can easily be added to the circuit, just by adding more input resistors. The general form of the R5 equation is:

$$R_5 = R_1/n$$

where *n* is the number of input resistors in the circuit.

The parts list for this project suggests a value of 10K (10,000 ohms) for resistors R1 through R4. To solve for resistor R5, just take one third of this value:

$$R_5 = 10,000$$
$$= 3333.333 \text{ ohms}$$

A standard 3.3K (3,000 ohms) resistor will be close enough in this application.

The output voltage from this circuit will be equal to the sum of the input voltages. To give you an idea of how this circuit functions, following is a list of several typical inputs and outputs:

Inputs			Output
V_1	V_2	V_3	V_{out}
1	1	1	3
1	−1	1	1
−1	−1	−1	−3
1.15	2.36	1.73	5.24
3.69	4.17	−2.54	5.32

There is one thing that is important to watch out for with this circuit. The total sum must not exceed the supply voltage of the op amps. If this happens, the output voltage will be clipped. For example, let's suppose a ±9-volt power supply is being used, and the following input voltages are applied to the circuit:

$$V_1 = 6 \text{ volts}$$
$$V_2 = 4 \text{ volts}$$
$$V_3 = 5 \text{ volts}$$

There is nothing inherently wrong with any of these input voltages individually, but their sum adds up to 15 volts. An op amp's output can never put out more than the supply voltage, so the output signal will be clipped to just a little under 9 volts, giving an incorrect result.

You might want to experiment with alternate resistor values in this circuit. Try changing the value of the feedback resistor (R4) to see the effects of nonunity gains in this circuit. Try both increasing (for greater than unity gain) and decreasing (for attenuation) this resistance.

Another possible area of experimentation is to try using different resistance values to the three input resistors (R1 through R3). This weights the inputs.

For simple experimentation, you probably won't have to worry about the value of resistor R5. Just keep this resistor's value in the right ballpark, and you should be fine, although there may be some minor errors in the output voltages. For a practical, finished version of this project, however, it is a good idea to calculate the appropriate value for resistor R5 to achieve the best, and most reliable results from the circuit.

ABSOLUTE VALUE CIRCUIT

In a number of practical applications, the magnitude of a voltage may be important, but not the polarity. For instance, a control circuit may need to detect 5-volt signals of either polarity (positive or negative). This could be done by using two separate detector circuits, one for +5 volts and the other for −5 volts. Besides being an inefficient solution, this may not always be practical.

What is needed in such a case is a circuit that extracts the *absolute value* of the input voltage. The absolute value is a concept from algebra, used to delete positive and negative signs. Essentially, the polarity sign is discarded and ignored. The absolute value of any number (or voltage) is always positive.

Absolute value circuit

Mathematically, absolute value is written as two bars on either side of the number to be treated:

$$|X| = \text{absolute value of } X$$

The absolute value function always returns a positive result, regardless of the polarity of the original number (or signal). The concept is made much clearer by looking at a few examples:

$$|5| = 5$$
$$|-7| = 7$$
$$|4.32| = 4.32$$
$$|-0.1| = 0.1$$
$$|19| = 19$$
$$|-19| = 19$$
$$-|3| = -3$$

Did that last one catch you off guard? If the sign is outside the bars, it is not part of the absolute value function.

A practical circuit for extracting the absolute value of a voltage signal is shown in Fig. 2-2. A suitable parts list for this project is given in Table 2-2.

In this circuit, IC1A passes and inverts any positive input voltages, while the diodes block out any negative voltage inputs. IC1B serves as an inverting summing amplifier. The original input signal to the circuit is fed into this op amp through resistor R5 and summed with the output of IC1A (through resistor R4). The voltage signal through resistor R5 is made one half the value of the voltage through resistor R4. These is accomplished by giving R4 a resistance half that of R5.

Table 2-2. Parts List for
Project 2: Absolute Value Circuit.

Part	Description
IC1	747 dual op amp
D1, D2	diode (1N4148, 1N914, or similar)
R1, R2, R5	22K 5% 1/4-watt resistor
R3, R4	10K 5% 1/4-watt resistor (see text)
R6	42K 5% 1/4-watt resistor
R7	4.7K 5% 1/4-watt resistor

Fig. 2-2 *Project 2: absolute value circuit.*

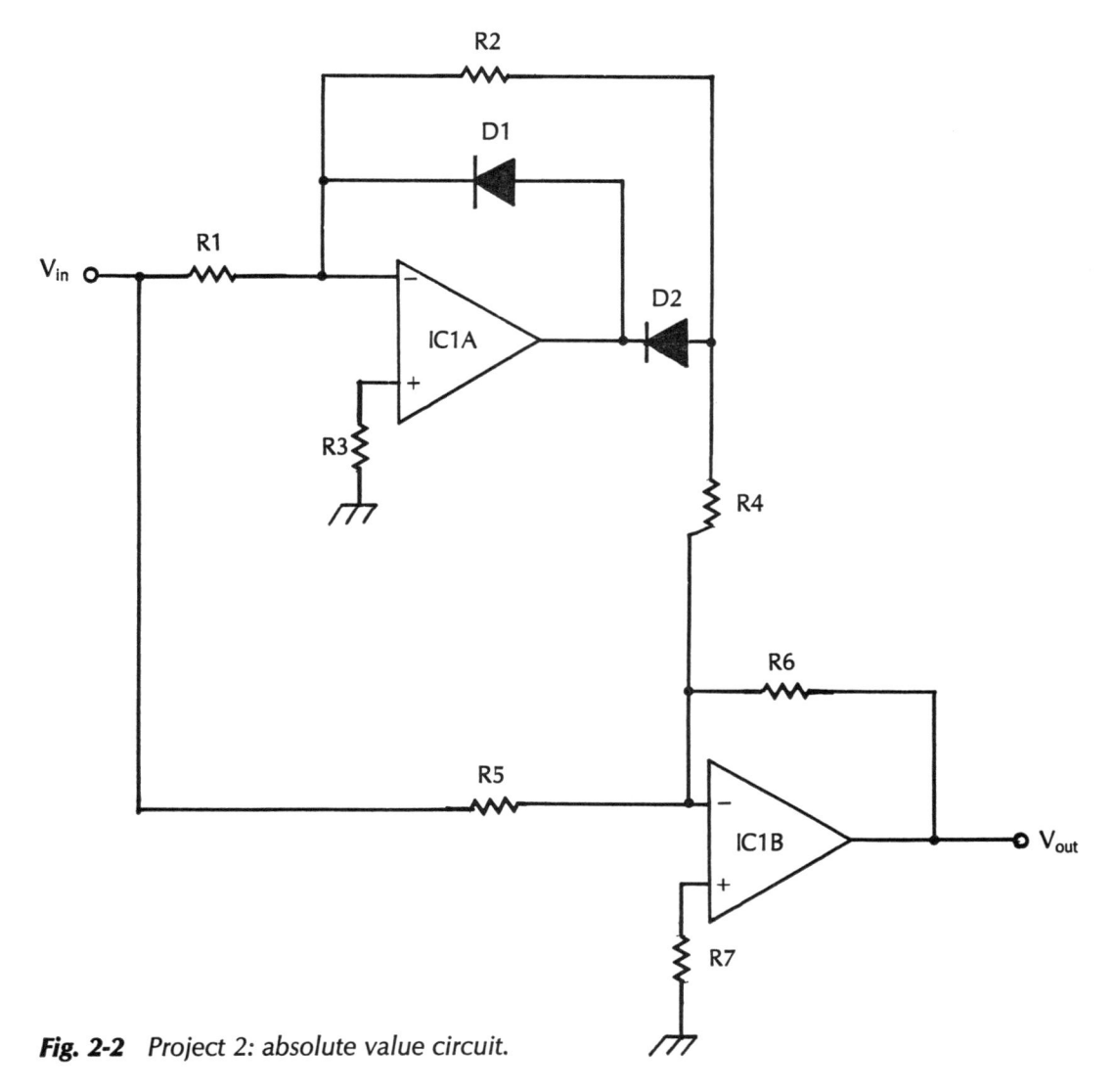

The parts list suggests a 10K resistor for R4 and a 22K resistor for R5. The value of R4 is not quite half that of R5 in this case. It should be close enough for most purposes. The resistor tolerances easily could introduce as much error. For more precise results, use low-tolerance resistors and use a 10K resistor in series with a 1K resistor for R4. This makes up 11K, which is half of 22K.

This example shows what happens with a couple of typical input signals. (For convenience, ignore the exact voltage drops through resistors R4 and R5 because they won't make any real difference in the final output signal.) Assume that the input voltage is +5 volts. IC1A passes and inverts this signal. Because this stage is set up for unity gain ($R_1 = R_2$), the output of IC1A equals −5 volts. We'll call this voltage V_a.

Voltage V_a is then fed into the IC1B inverting input through R4, along with half the original input signal through R5. These two voltages are then summed and inverted by IC1B. This stage is set up for a gain of two:

$$\begin{aligned} V_{out} &= -2 \times (V_{in}/2 + V_a) \\ &= -2 \times (5/2 + -5) \\ &= -2 \times (2.5 - 5) \\ &= -2 \times -2.5 \\ &= 5 \text{ volts} \end{aligned}$$

(Multiplying a negative number by a negative number always results in a positive number.)

Now, consider what happens when the input voltage is negative, say −4 volts. This signal will be blocked by IC1A and the diodes, so V_a has a value of 0 in this case. The output voltage works out to a value of:

$$\begin{aligned} V_{out} &= -2 \times (-4/2 + 0) \\ &= -2 \times -2 \\ &= 4 \text{ volts} \end{aligned}$$

The output of this circuit will always be positive, regardless of the polarity of the original input signal. The magnitude of the signal is not altered by the circuit.

For this circuit to work properly, the following value relationships must be maintained between the resistors:

$$\begin{aligned} R_1 &= R_2 = R_5 \\ R_4 &= R_1/2 \\ R_3 &= R_4 \\ R_6 &= 2R_5 \\ R_7 &= R_1/4 \end{aligned}$$

For applications that require very precise voltage matching, use a 50K trimpot for R6. This will permit you to fine tune the circuit's output exactly to the correct value, compensating for any other inaccuracies within the circuitry.

Absolute value circuits are useful in analog computation systems, and detection and automation applications. They can also be used for precision full-wave rectification.

WINDOW COMPARATOR

A *comparator* is a circuit that looks at an unknown voltage and compares it with a known reference voltage. A simple comparator circuit simply indicates whether or not the unknown input voltage is greater than or less than a single constant reference voltage. More sophisticated comparator circuits determine whether or not the unknown input voltage is within a specific range of voltages.

One type of specialized comparator circuit is the *window comparator*. A circuit of this type is shown in Fig. 2-3. A window comparator's output indicates whether or not the unknown input voltage is within a specific range, set by a pair of reference voltages.

If the unknown input voltage is within the specified band of reference voltages, the circuit's output will be zero. If the unknown input voltage is outside the specified band, the circuit output will be saturated at a high voltage (a little less than the supply voltage). The polarity of the saturation voltage indicates if the unknown input voltage is too high or too low. If the unknown input voltage is higher than the upper reference voltage limit, the output of the circuit will be at positive saturation (V+). Similarly, if the unknown input voltage is less than the lower reference voltage limit, the circuit's output will be at negative saturation (V−).

The real key to the functioning of this circuit is in the output diodes D1 and D2. Almost any standard, small signal silicon diodes can be used in this circuit. These diodes are what make the circuit act as a threshold detector, or window comparator, as described above. IC1A and diode D1 handle positive input signals, and IC1B and diode D2 are set up to handle negative input signals.

The specified band of reference voltages is determined by a pair of simple resistive voltage dividers. The upper reference voltage is set by the ratio of the values of resistors R1 and R2. Similarly, the lower reference voltage is controlled by the relative values of resistors R4 and R5.

IC1A compares the unknown input voltage with the upper

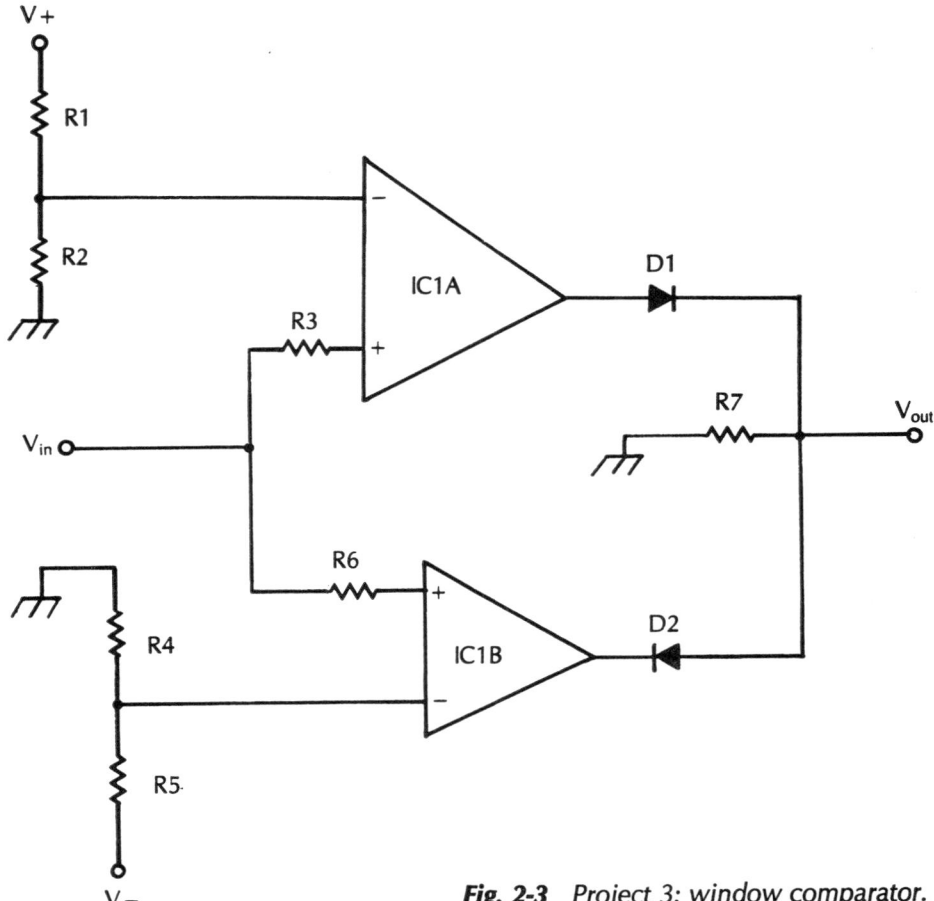

Fig. 2-3 Project 3: window comparator.

reference voltage, and IC1B compares it with the lower reference voltage. For symmetrical operation of the circuit, resistors R3 and R6 should have identical values.

In the following example, assume the circuit is being operated with a ±12-volt power supply. The suggested component values from the parts list (Table 2-3) will be used.

Finding the reference voltages in this comparator circuit is simply a matter of using Ohm's Law. The total resistance in the upper voltage divider string is just the sum of the values of resistors R1 and R2:

$$R_u = R_1 + R_2$$
$$= 22,000 + 10,000$$
$$= 32,000 \text{ ohms}$$

Table 2-3. Parts List for Project 3: Window Comparator.

Part	Description
IC1	747 dual op amp
D1, D2	diode (1N914, 1N4148, or similar)
R1, R5	22K 5% 1/4-watt resistor
R2, R4, R7	10K 5% 1/4-watt resistor
R3, R6	6.8K 5% 1/4-watt resistor

Ohm's Law tells us the current flowing through these two resistors:

$$I = E/R$$
$$= 12/32{,}000$$
$$= 0.000375 \text{ ampere}$$
$$= 0.375 \text{ mA}$$

The upper reference voltage will be equal to the voltage drop across resistor R2. In this example, this works out to:

$$E = IR$$
$$= 0.000375 \times 10{,}000$$
$$= 3.75 \text{ volts}$$

According to the parts list, the two voltage divider networks are equal. That is:

$$R_5 = R_1$$
$$R_4 = R_2$$

As a result, the lower reference voltage is the same as the upper voltage, except the polarity is reversed. In the example, the lower reference voltage is −3.75 volts.

This equality between the upper and lower reference voltages is not at all necessary. It is just as easy to set up this circuit for a reference window that is symmetrical around zero.

Using the component values outlined in the parts list, if the unknown input voltage is more negative than −3.75 volts, the output will be about −12 volts. If the unknown input voltage is greater than +3.75 volts, the output will be just under +12 volts. The output will be 0 if the unknown input voltage is between −3.75 volts and +3.75 volts, because the diodes will inhibit output from the op amps. Here are some typical examples:

Input Voltage	Output at IC1A	Output at D1	Output at IC1B	Output at D2	Sum of D1 + D2 Output at V_{out}
5	+12	+12	+12	0	12
4	+12	+12	+12	0	12
3	−12	0	+12	0	0
0	−12	0	+12	0	0
−3	−12	0	+12	0	0
−4	−12	0	−12	−12	−12
−5	−12	0	−12	−12	−12

and so forth. A window comparator circuit can come in handy for many testing, automation, and monitoring applications. The operation of this window comparator circuit is illustrated in the graph of Fig. 2-4.

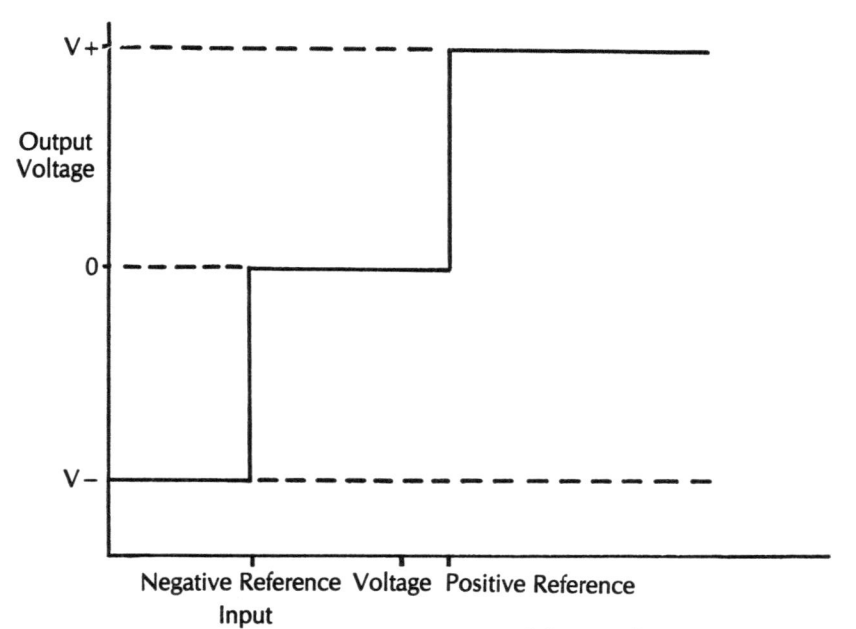

Fig. 2-4 This graph illustrates the operation of the window comparator circuit of Fig. 2-3.

MAGNITUDE DETECTOR

A close relative of the window comparator of the preceding project is the *magnitude comparator,* shown in Fig. 2-5. Notice how

Fig. 2-5 Project 4: magnitude detector.

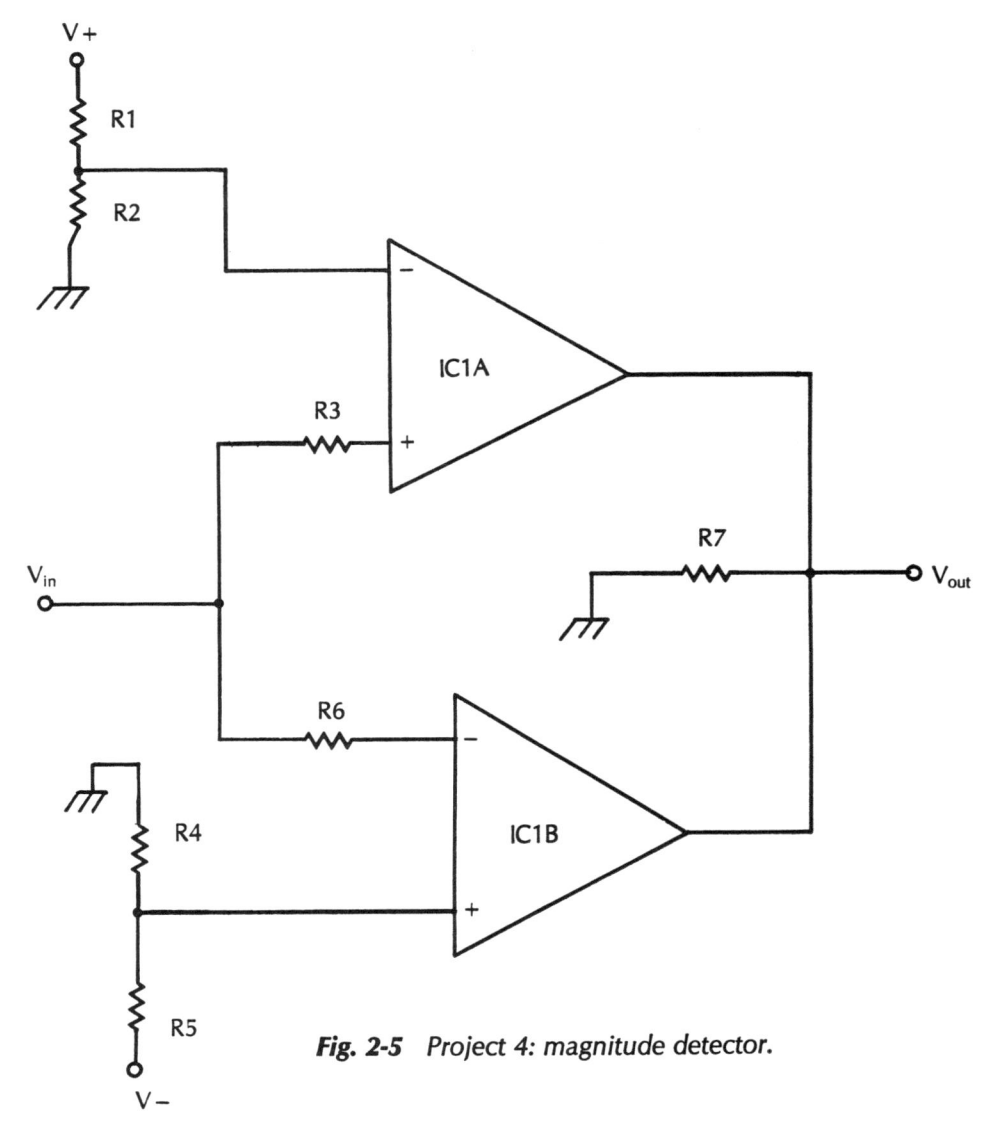

much this circuit resembles the circuit of Fig. 2-3. The main difference is the absence of the two output diodes, and the reversal of the inputs to IC1B.

The magnitude detector has two output stages. One output stage handles input voltages within the specified range, and the other deals with out-of-range signals. The window range is set by the two voltage divider networks (R1/R2 and R5/R4), just as in the preceeding project.

As in the earlier project, IC1A compares the unknown input voltage with the upper reference voltage, and IC1B compares it with the lower reference voltage. For symmetrical operation of the circuit, resistors R3 and R6 should have identical values.

In the following example, assume the circuit is being operated off of a ±12-volt power supply. The suggested component values from the parts list will be used.

Finding the reference voltages in this comparator circuit is simply a matter of using Ohm's Law. The total resistance in the upper voltage divider string is just the sum of the values of resistors R1 and R2:

$$\begin{aligned} R_u &= R_1 + R_2 \\ &= 22{,}000 + 10{,}000 \\ &= 32{,}000 \text{ ohms} \end{aligned}$$

Ohm's Law tells us the current flowing through these two resistors:

$$\begin{aligned} I &= E/R \\ &= 12/32{,}000 \\ &= 0.000375 \text{ ampere} \\ &= 0.375 \text{ mA} \end{aligned}$$

The upper reference voltage will be equal to the voltage drop across resistor R2. In this example, this works out to:

$$\begin{aligned} E &= IR \\ &= 0.000375 \times 10{,}000 \\ &= 3.75 \text{ volts} \end{aligned}$$

According to the parts list, the two voltage divider networks are equal. That is:

$$\begin{aligned} R_5 &= R_1 \\ R_4 &= R_2 \end{aligned}$$

As a result, the lower reference voltage is the same as the upper voltage, except the polarity is reversed. In our example, the lower reference voltage is −3.75 volts. This equality between the upper and lower reference voltages is not at all necessary. It is just as easy to set up this circuit for a reference window that is symmetrical around zero. Experiment with alternate component values in this circuit. Try different resistance ratios in the voltage dividers.

30 Operational circuits

Table 2-4. Parts List for Project 4: Magnitude Detector.

Part	Description
IC1	747 dual op amp
R1, R5	22K 5% 1/4-watt resistor
R2, R4, R7	10K 5% 1/4-watt resistor
R3, R6	6.8K 5% 1/4-watt resistor

When the unknown input voltage is within the specified range, both IC1B's output and IC1A's output go into negative saturation. These two signals keep the circuit's output at negative saturation.

Now, let's assume that the unknown input voltage becomes more positive and becomes greater than the upper reference voltage (+3.75 volts). This will cause IC1A to go into positive saturation while IC1B stays at negative saturation so they cancel each other out, driving the output to zero. The opposite same thing happens when the unknown input voltage drops below the lower reference voltage (−3.75 volts).

As long as the unknown input voltage is within the specified range, the circuit output remains low. If the unknown input voltage goes outside the band of interest in either direction (either positively, or negatively) the op amps are forced to cancel each other, so the circuit output goes to zero. Following are some typical examples:

Input Voltage	Output at IC1A	Output at IC1B	Output at V_{out}
5	12	−12	0
4	12	−12	0
3	−12	−12	−12
2	−12	−12	−12
1	−12	−12	−12
0	−12	−12	−12
−1	−12	−12	−12
−2	−12	−12	−12
−3	−12	−12	−12
−4	−12	12	0
−5	−12	12	0

The operation of this magnitude comparator circuit is illustrated in Fig. 2-6. If the inputs to both comparators were

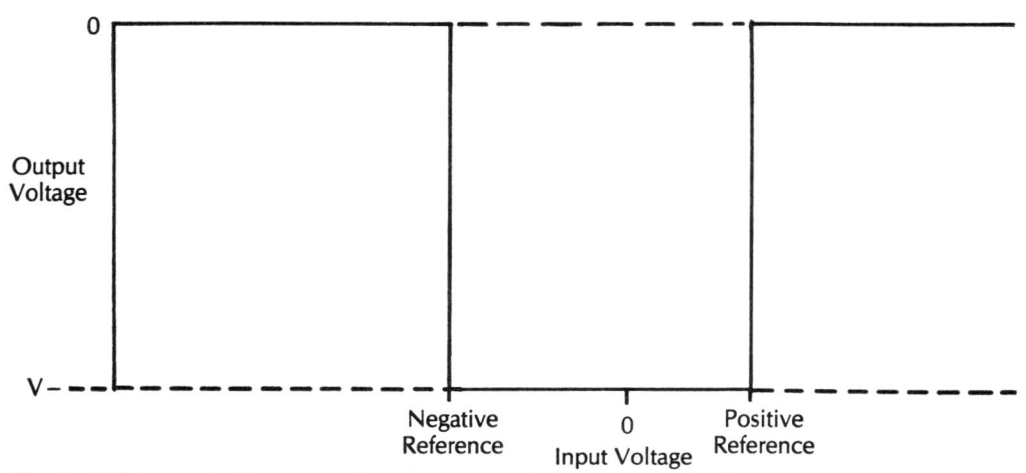

Fig. 2-6 *This graph illustrates the operation of the magnitude detector circuit of Fig. 2-5.*

switched, then when the input voltage was within the band, the output would be positive instead of negative.

PEAK DETECTOR

In some applications it may be necessary to electronically determine the maximum value of a varying voltage within a given period of time. This can be done with a circuit known as a *peak detector*. The simplest peak detector circuit consists of just a diode and a capacitor, as illustrated in Fig. 2-7.

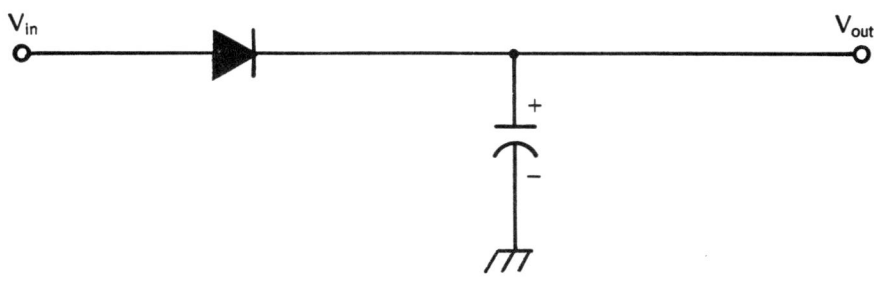

Fig. 2-7 *This is a very simplified peak detector circuit.*

The diode allows the voltage to flow in only one direction, from the input to the capacitor and on to the output. As the input voltage is increased, the capacitor is charged up through the diode. If the input voltage begins to drop off the capacitor will

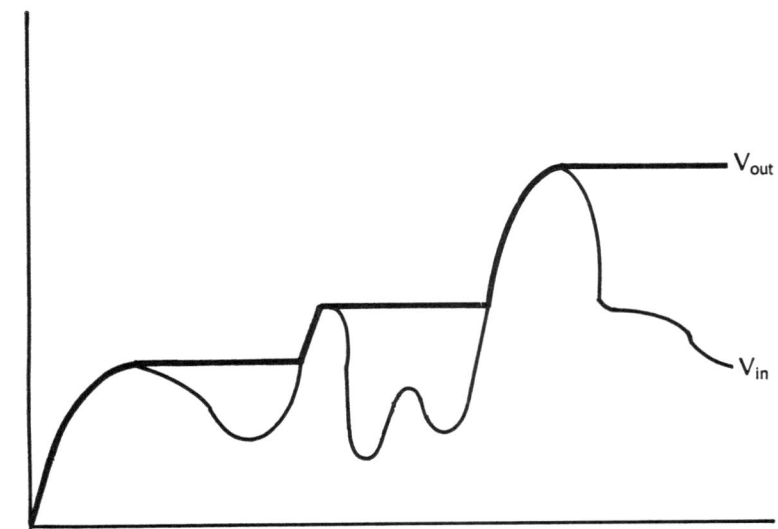

Fig. 2-8 *This graph illustrates the operation of a typical peak detector circuit.*

not discharge because that would reverse-bias the diode. The capacitor holds its peak charged value. If the input voltage begins to increase again, and exceeds the previous peak value, the capacitor will charge further.

The action of this type of circuit is illustrated in the graph of Fig. 2-8. The input voltage is tracked and followed until a maximum level is reached. This peak voltage value is held at the output by the peak detector circuit.

This simplified, idealized peak detector circuit isn't very practical. The capacitor soon will discharge through the load circuit. Even without a load, a capacitor can not hold a charge indefinitely. The stored voltage will begin to taper off, due to capacitor leakage.

A practical peak detector circuit is shown in Fig. 2-9. A suitable parts list for this project appears as Table 2-5. The first op amp in this circuit (IC1A) functions as a precision diode, essentially taking the place of the simple diode in Fig. 2-7. The second op amp stage (IC1B) is a noninverting buffer amplifier (unity gain) to isolate the storage capacitor from the load. This minimizes leakage problems due to the capacitor discharging prematurely through the load circuit.

Fig. 2-9 *Project 5: peak detector.*

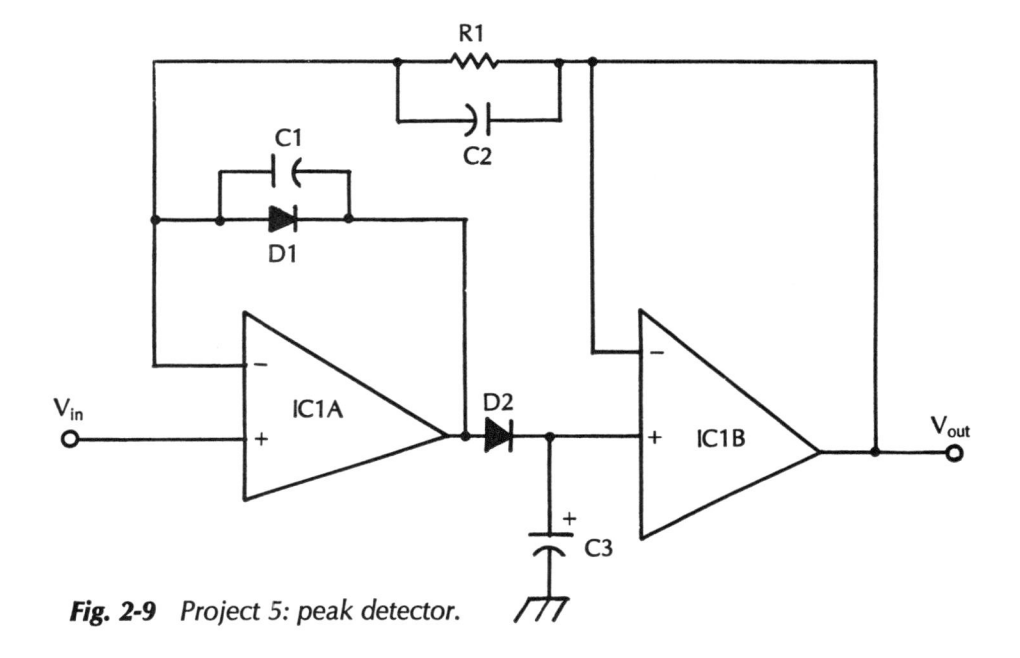

Table 2-5. Parts List for Project 5: Peak Detector.

Part	Description
IC1	747 dual op amp
D1, D2	diode (1N4148, 1N914, or similar)
C1, C2	0.1 μF capacitor
C3	10 μF 50-volt tantalum capacitor (see text)
R1	10K 5% 1/4-watt resistor

Capacitor C3 serves as the actual peak holding capacitor in this circuit. It should have a relatively large value. In a circuit of this type it is strongly advised that you use a high-grade low-leakage capacitor. A tantalum capacitor probably would be a good choice in this application. An inexpensive electrolytic capacitor won't give very satisfactory results in a peak detector circuit.

The other two capacitors in this circuit (C1 and C2) are included to improve the circuit stability and to prevent overshoot from a stepped input signal, which can be a problem with some

34 Operational circuits

peak detector circuits. These capacitors are not always needed, but it is a good idea to include them, just in case. They are cheap insurance against possible problems. None of the component values are particularly critical in this circuit.

One major advantage of this particular peak detector circuit is that IC1A cannot latch up (get stuck at a certain output voltage), as can happen with some other peak detector circuits.

INVERTING PEAK DETECTOR

Another practical peak detector circuit is shown in Fig. 2-10. The primary functional difference between this circuit and the circuit of Fig. 2-6 is that this one inverts the polarity of the output voltage. Obviously, the output op amp in this circuit is operated in the inverting mode.

IC1A again serves as a precision diode, while IC1B tracks and inverts the signal. Capacitor C2 is the peak holding capacitor in this circuit. Positive feedback from the output of IC1B is returned to the noninverting input of IC1A through resistor R4, and capacitor C3. The parts list for this project is given in Table 2-6.

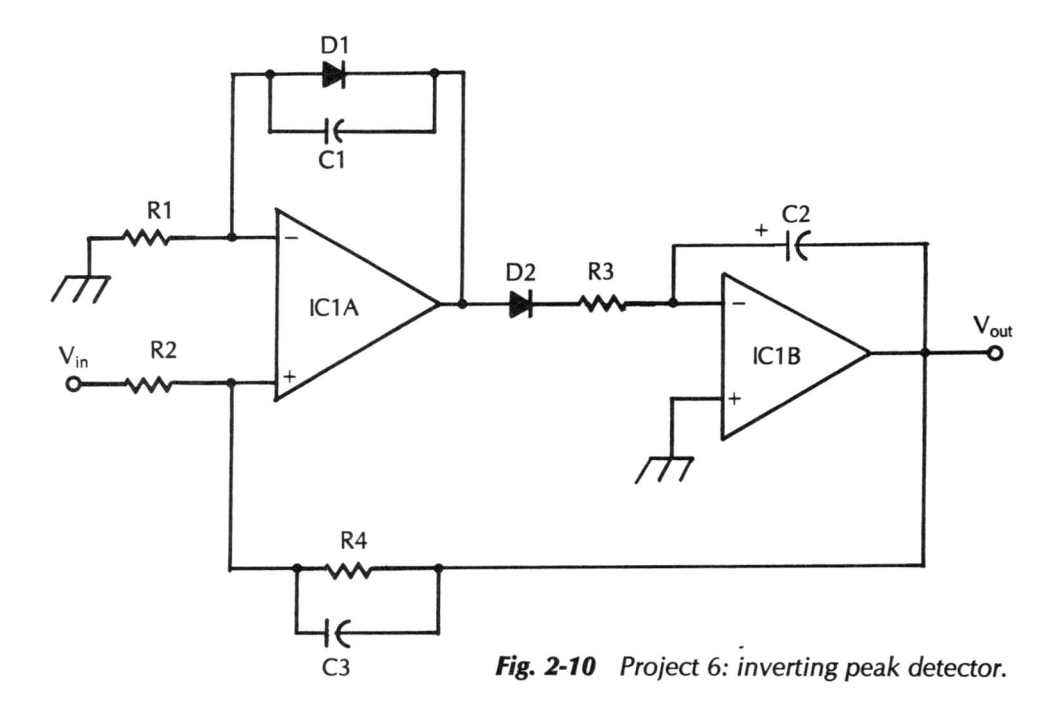

Fig. 2-10 Project 6: inverting peak detector.

**Table 2-6. Parts List for
Project 6: Inverting Peak Detector.**

Part	Description
IC1	747 dual op amp
D1, D2	diode (1N4148, 1N914, or similar)
C1, C3	0.1 µF capacitor
C2	10 µF 50-volt tantalum capacitor
R1	4.7K 5% 1/4-watt resistor
R2, R4	10K 5% 1/4-watt resistor
R3	22K 5% 1/4-watt resistor

To minimize any dc errors within this circuit, the value of resistor R1 should be equal to the parallel combination of resistors R2 and R4. That is:

$$R_1 = \frac{R_2 \times R_4}{R_2 + R_4}$$

By keeping the values of R2 and R4 equal, resistor R1 can be given one-half the value of R2 or R4.

PEAK DETECTOR WITH MANUAL RESET

The circuit shown in Fig. 2-11 is probably the simplest possible two op amp peak detector circuit. The parts list for this simple project is given in Table 2-7.

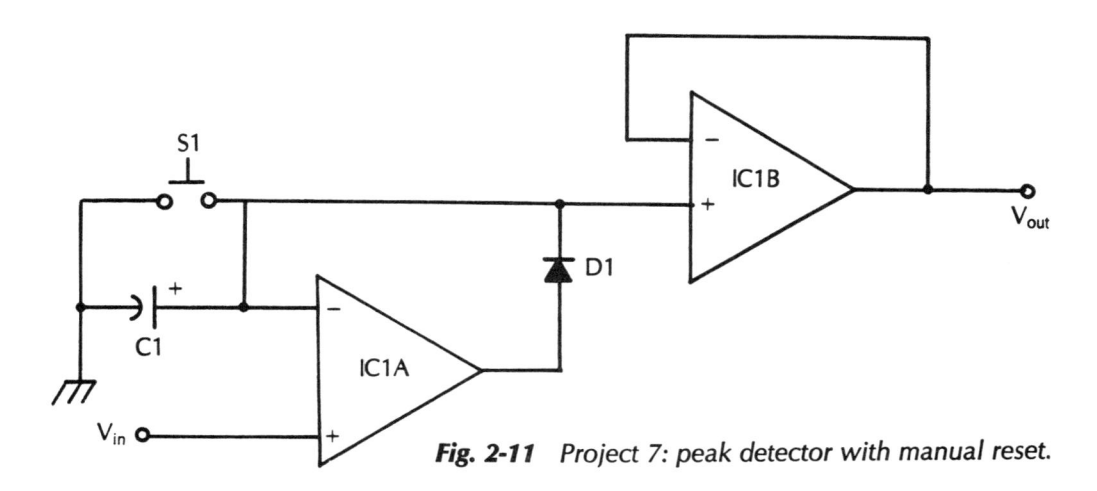

Fig. 2-11 Project 7: peak detector with manual reset.

Table 2-7. Parts List for Project 7:
Peak Detector with Manual Reset.

Part	Description
IC1	747 dual op amp
D1	diode (1N4148, 1N914, or similar)
C1	10 μF 50-volt tantalum capacitor
S1	normally open SPST push-button switch

Aside from the exceptionally low parts count, the chief advantage of this circuit is that it is manually resettable. Momentarily closing switch S1 allows the holding capacitor to discharge and resets the circuit's output to zero. This simple manual reset can be added to almost any peak detector circuit.

To briefly examine just what happens within this circuit during operation, assume that the output has just been reset to zero. Now apply a positive voltage to the circuit's input (the noninverting input of IC1A). Since this input signal is not inverted at the op amp's output, diode D1 is forward biased. Even a very small input voltage will forward-bias the diode because of the high gain of the open loop op amp. The effects of the small voltage drop across the diode are almost negligible in this circuit. For all intents and purposes, you can ignore the diode voltage drop.

Since the diode is forward biased, it allows current to flow back into holding capacitor C1. This capacitor begins to charge up. When the charge on C1 equals the input voltage, the differential voltage across the inputs of op amp IC1A is 0. The output of this op amp now drops to 0. No more current flows through the diode, so the capacitor can't charge up any further. It can't discharge, because that would reverse bias the diode. The diode blocks current flow in that direction. The capacitor must hold its charge.

If the input voltage is now increased to a higher (more positive) value, the differential input voltage of IC1A will go above zero. The output of the op amp will forward-bias diode D1 again, and the capacitor will be charged up to this new maximum value. If the input voltage decreases, diode D1 will be reverse biased, so the capacitor will continue to hold its charge at the last peak value.

The voltage across the holding capacitor (C1) will always be equal to the largest positive voltage that has been fed into the circuit since the last time it was reset. Closing switch S1 shorts out the holding capacitor, permitting it to very quickly discharge to ground, resetting the circuit.

IC1B is simply a noninverting voltage follower to act as an output buffer for the circuit. This buffer limits loading effects on the capacitor. The capacitor sees a very high impedance at the input of IC1B, so the capacitor can not discharge appreciably through the buffer stage.

With a good, high-quality capacitor, the peak voltage can be held for periods lasting up to several minutes. For the best results, use a low-leakage device, such as a tantalum capacitor. An electrolytic capacitor will work in some applications, but the circuit's operation will tend to be somewhat sluggish, and the peak value will tend to drift back toward 0 if it is held for more than a few seconds. But remember, no electronics component is ever perfect. Any capacitor will have some degree of internal leakage, that will cause its charge to eventually drift off towards zero. The capacitor will also suffer some minimal discharging through the leakage current of the diode, and the bias current required by the op amps. The output buffer will place some slight loading on the the holding capacitor, although the loading effects will be considerably less than would be experienced with most external load circuits.

LOGARITHMS

After addition and subtraction, the two most basic mathematical operations are multiplication and division. Unfortunately, these operations are somewhat difficult to perform electronically—at least directly. Therefore, you must take a brief detour and explore the concept of logarithms. You'll see why shortly.

The *logarithm* of a number is the exponent that indicates the power to which a base number must be raised to produce a given value. Common logarithms use a base of 10. Following are a few numbers and their logarithms:

Number	Logarithm
1	0
2	0.30103
3	0.4771213

Operational circuits

Number	Logarithm
4	0.60206
5	0.69897
10	1
12	1.0791812
20	1.30103
25	1.39794
30	1.4771213
100	2
1000	3
10000	5

Raising 10 to the logarithm produces the original number:

$$10^0 = 1$$
$$10^{0.30103} = 2$$
$$10^{0.4771213} = 3$$
$$10^1 = 10$$
$$10^2 = 100$$
$$10^3 = 1000$$

Raising any number to the power of zero (X^0) always results in 1, so the logarithm of 1 is 0, no matter what base is used. The logarithm of zero is infinity in the negative direction. Negative logarithm values indicate numbers less than 1:

$$\text{Log } 0.2 = -0.69897$$
$$\text{Log } 0.01 = -1$$
$$\text{Log } 0.005 = -2.30103$$

To work from the logarithm back to the original number is called finding the antilogarithm. For example:

$$\text{Antilog } 0.30103 = 2$$
$$\text{Antilog } 1 = 10$$
$$\text{Antilog } 2 = 100$$

Well, so what? What are logarithms good for, and why are you looking at them here? Logarithms can come in extremely handy in many mathematical applications. In fact, old-fashioned slide rules (back in the precalculator days) used the principle of logarithms to solve various equations. Logarithms can be used to perform multiplication and division, and convert linear values in decibels (dB), among many other applications.

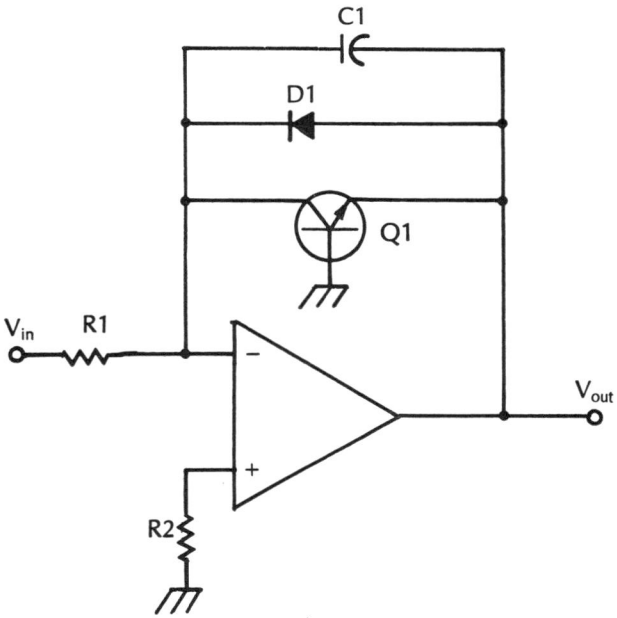

Fig. 2-12 *This circuit converts a linear input voltage into a logarithmic output voltage.*

A logarithmic amplifier built around an op amp uses a resistor as the input component, and a transistor as the primary feedback component. The base of the transistor is grounded. The basic circuit of a logarithmic amplifier is shown in Fig. 2-12.

The collector/emitter curve of many transistors is very close to a true logarithmic curve, making the devices ideal for use in logarithmic amplifier circuits. The collector/emitter current through the transistor will be equal to the input current, thanks to the nature of the op amp. As a result, the output voltage will be proportional to the logarithm of the input voltage. A power transistor is often used in such applications to reduce the series feedback resistance, and improve the circuit's current handling capabilities. However, the specific characteristics of the transistor used generally aren't too critical. The capacitor and the diode in this circuit help smooth out the logarithmic curve.

The opposite of a logarithm is the antilogarithm. An antilogarithmic amplifier can be created easily, by reversing the input and output components of a logarithmic amplifier, as shown in Fig. 2-13.

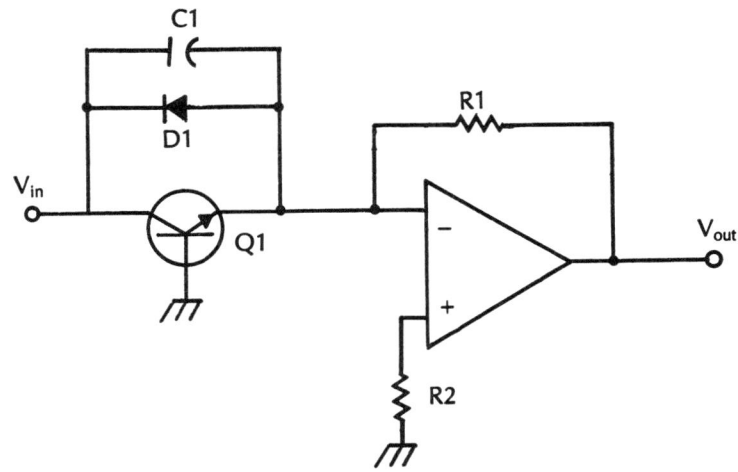

Fig. 2-13 *The antilogarithmic amplifier is just the opposite of the logarithmic amplifier circuit of Fig. 2-12.*

MULTIPLICATION CIRCUIT

Multiplying an input signal by a constant value is not particularly difficult. All you need is a noninverting amplifier circuit with the appropriate amount of gain. The output voltage of a noninverting amplifier equals the input voltage multiplied by the circuit gain:

$$V_{out} = V_{in} \times G$$

Therefore, if you need a circuit to multiply the input signal by a constant factor of 10, you need only design a noninverting amplifier with a gain of 10.

This is functional, but severely limited, and hopelessly inadequate in many applications. A little more versatility can be achieved by using a potentiometer as the feedback resistor. In this way, the multiplication factor can be manually adjusted, although calibration might be rather difficult.

In many applications, it might be necessary to multiply two variable input voltages. In this case, an entirely different approach may be used. This is where the logarithms and antilogarithms discussed in the last few pages come in. Logarithms exhibit many interesting characteristics. For instance, if you add the logarithms of two numbers and then take the antilogarithm of their sum, you will get the product of the two original numbers.

In algebraic terms, this idea is expressed as the following formula:

$$\text{Antilog}(\text{Log } A + \text{Log } B) = A \times B$$

The use of logarithms permits multiplication by simple addition. To prove this works, try out a simple example. Say you need to multiply 2 times 3. Of course, you already know the answer should work out to 6. See if the logarithm method will get the desired result:

$$\begin{aligned} C &= A \times B \\ &= \text{Antilog}(\text{Log } A + \text{Log } B) \\ &= \text{Antilog}(\text{Log } 2 + \text{Log } 3) \\ &= \text{Antilog}(0.30103 + 0.4771213) \\ &= \text{Antilog } 0.7781513 \\ &= 6 \end{aligned}$$

Yes, you got the correct result.

This was a very trivial example. The logarithms probably seemed like more trouble than they're worth. The logarithmic method is more useful in complex problems, such as:

$$\begin{aligned} C &= 3607.814 \times 90011.0043 \\ &= \text{Antilog}(\text{Log } 3607.814 + \text{Log } 90011.0043) \\ &= \text{Antilog}(3.5572441 + 4.9542956) \\ &= \text{Antilog } 8.5115397 \\ &= 324742960.21 \end{aligned}$$

You can check this result with a calculator or a computer.

You can combine five basic op amps to perform this logarithmic multiplication operation for you. You need two logarithmic amplifiers, a summing amplifier, an inverting voltage follower, and an antilogarithmic amplifier. The block diagram for a complete op amp multiplier circuit is shown in Fig. 2-14.

The schematic diagram for the multiplier circuit is illustrated in Fig. 2-15. A suitable parts list for this project is given in Table 2-8.

DIVIDER CIRCUIT

Division can be performed with logarithms in a manner similar to multiplication. Instead of adding logarithms of two original numbers and taking the antilogarithm, subtract one logarithm

42 Operational circuits

Fig. 2-14 This is the block diagram of a multiplication circuit.

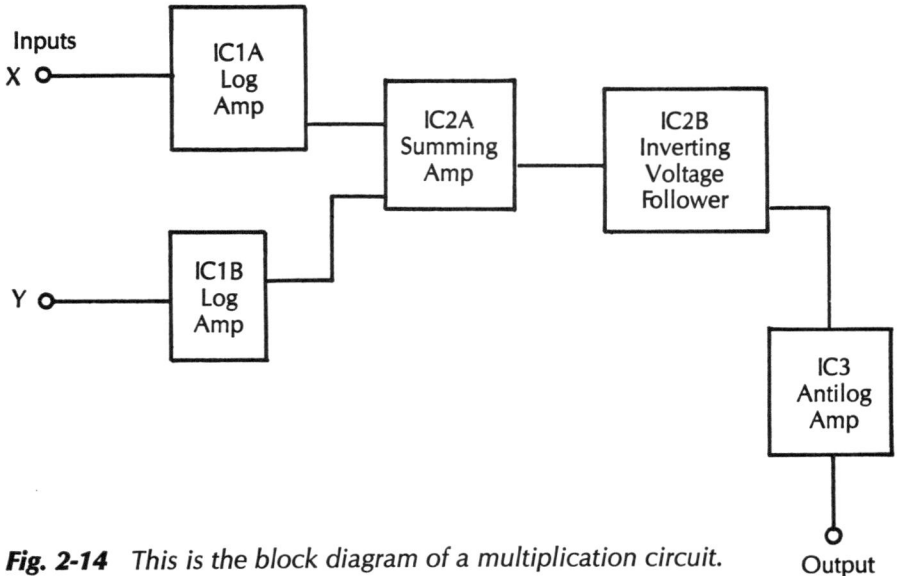

Table 2-8. Parts List for Project 8: Multiplication Circuit.

Part	Description
IC1, IC2	747 dual op amp
IC3	single op amp (see text)
Q1, Q2, Q3	npn transistor (2N3904, or similar)
D1, D2, D3	diode (1N4001, or similar)
C1, C2, C3	0.1 µF capacitor
R1, R2, R5, R6, R11, R12	47K 1/4-watt 5% resistor
R3, R7	2.2K 1/4-watt 5% resistor
R4, R8, R9, R10	10K 1/4-watt 5% resistor

from the other, then take the antilogarithm of the difference:

$$A/B = \text{Antilog} (\text{Log } A - \text{Log } B)$$

As an example, divide 15 by 3. The result should be 5, of course:

$$\begin{aligned}
C &= 15/3 \\
&= \text{Antilog} (\text{Log } 15 - \text{Log } 3) \\
&= \text{Antilog} (1.1760913 - 0.4771213) \\
&= \text{Antilog } 0.69897 \\
&= 5
\end{aligned}$$

Divider circuit 43

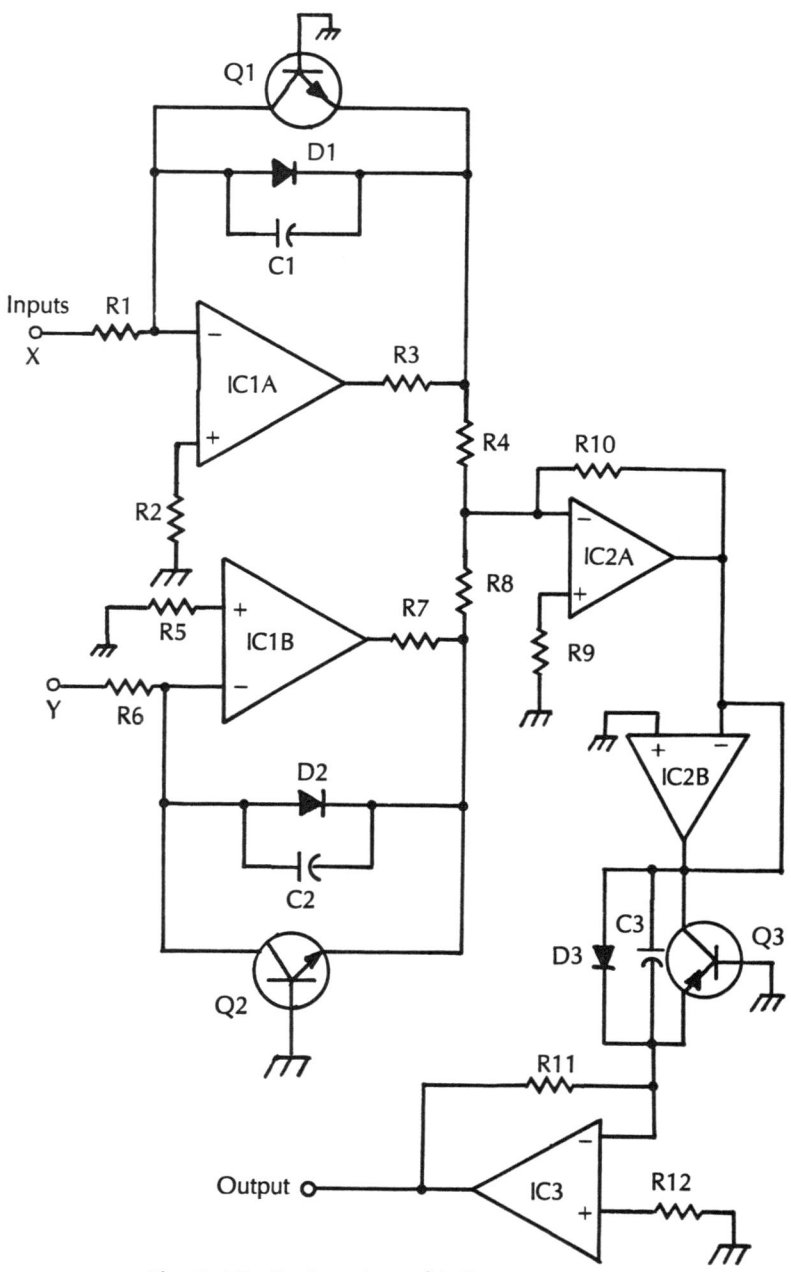

Fig. 2-15 Project 8: multiplication circuit.

44 Operational circuits

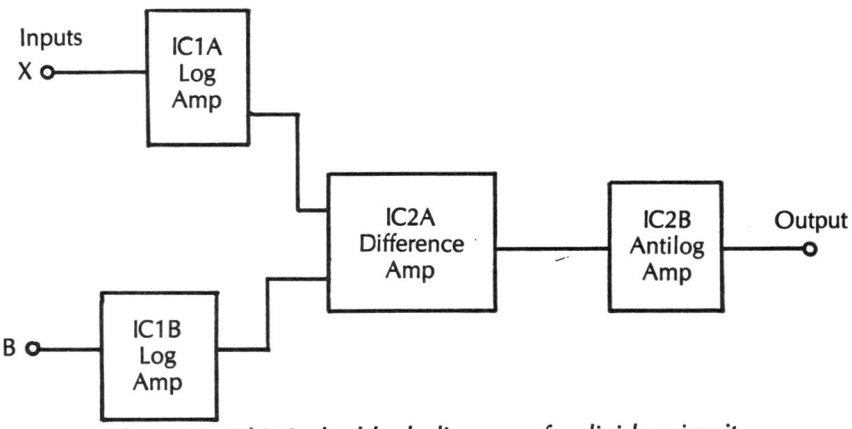

Fig. 2-16 *This is the block diagram of a divider circuit.*

Yes, the logarithmic method gave you the correct result.

The block diagram for an op amp division circuit is shown in Fig. 2-16. The schematic diagram is illustrated in Fig. 2-17. A suitable parts list for this project appears as Table 2-9.

EXPONENT CIRCUIT

Sometimes it is necessary to raise an input signal to a specific power, or exponent. This is simply multiplying the number by itself a specific number of times. For example, 5 raised to the fourth power is:

$$5^4 = 5 \times 5 \times 5 \times 5$$
$$= 625$$

Table 2-9. Parts List for Project 9: Divider Circuit.

Part	Description
IC1, IC2	747 dual op amp
Q1, Q2, Q3	npn transistor (2N3904, or similar)
D1, D2, D3	diode (1N4001, or similar)
C1, C2, C3	0.1 μF capacitor
R1, R2, R3, R4, R8, R9	47K 1/4-watt 5% resistor
R5, R6, R7	10K 1/4-watt 5% resistor

Exponent circuit 45

Fig. 2-17 *Project 9: divider circuit.*

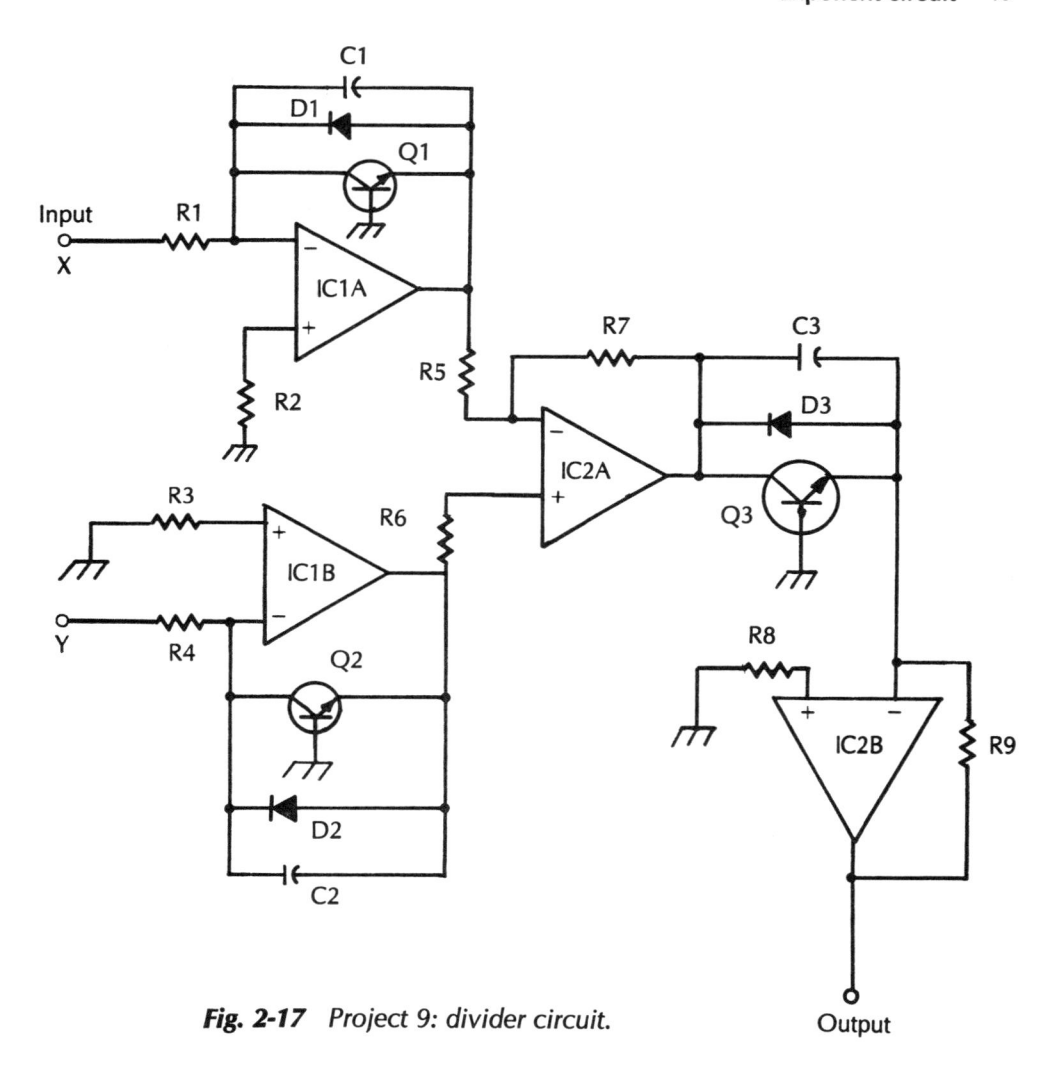

A logarithmic amplifier will again allow you to electronically perform this function fairly simply. To raise a number, A, to the nth power, use this logarithmic formula:

$$A = \text{Antilog}(\text{Log } A \times n)$$

Check this out by solving for 5 to the fourth power again:

$$\begin{align} C &= 5^4 \\ &= \text{Antilog}(\text{Log } 5 \times 4) \\ &= \text{Antilog}(0.69897 \times 4) \end{align}$$

46 Operational circuits

$$= \text{Antilog } 2.79588$$
$$= 625$$

The logarithmic method consistently gives the correct results.

Since the power (*n*) is likely to be a constant in many applications, you can use the gain of a noninverting amplifier to perform the multiplication. The block diagram for the exponent circuit is illustrated in Fig. 2-18. Notice that instead of a noninverting amplifier, there is an inverting amplifier, followed by an inverting voltage follower. This gives the same final results. With op amps, inverting amplifiers are usually more convenient to work with than noninverting amplifiers.

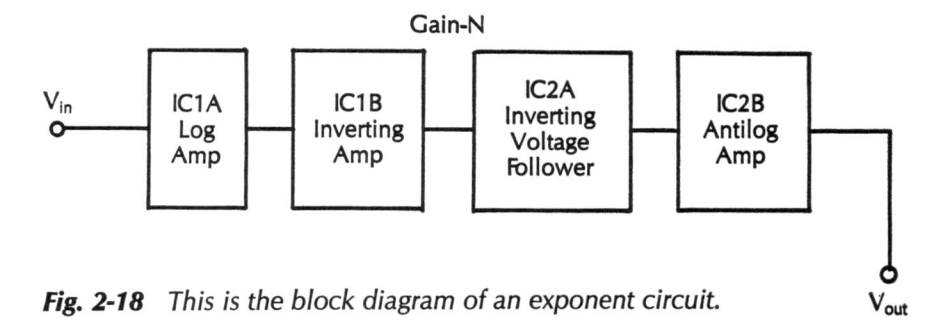

Fig. 2-18 *This is the block diagram of an exponent circuit.*

The complete schematic diagram for the exponent circuit is shown in Fig. 2-19. A suitable parts list for this project is given in Table 2-10. Select the value of resistor R4 to set the desired gain. The gain will be equal to:

$$G = R_4/R_3$$
$$= R_4/10000$$

Here are a few typical examples:

R4	G
10K	1
22K	2.2
33K	3.3
47K	4.7
100K	10
120K	12

Fig. 2-19 Project 10: exponent circuit.

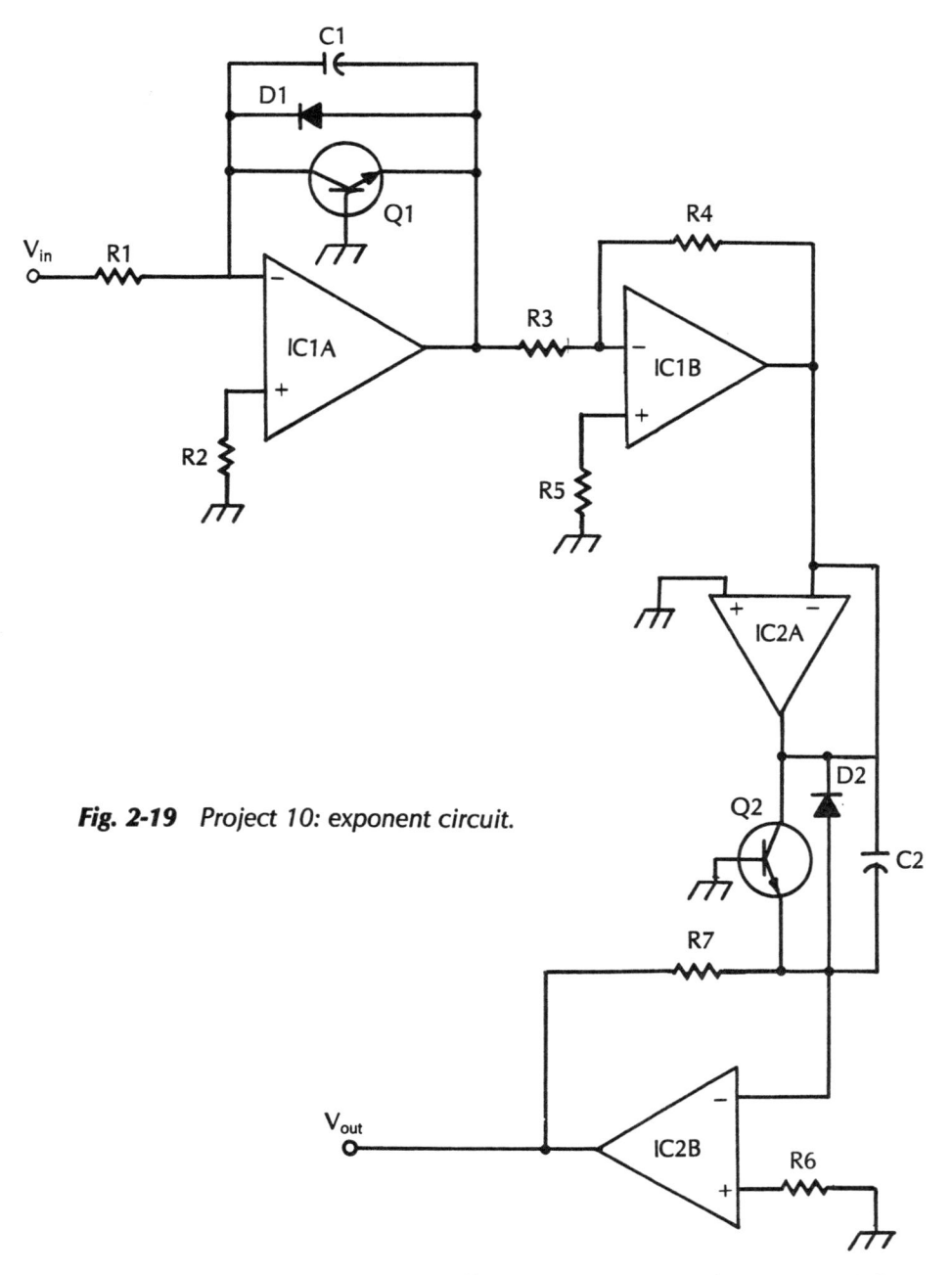

For greater flexibility, you could replace the fixed gain amplifier stage (IC1B and IC2A) with a two-input multiplication circuit, as shown back in Fig. 2-15.

Operational circuits

Table 2-10. Parts List for Project 10: Exponent Circuit.

Part	Description
IC1, IC2	747 dual op amp
Q1, Q2	npn transistor (2N3904, or similar)
D1, D2	diode (1N4001, or similar)
C1, C2	0.1 μF capacitor
R1, R2, R5, R6, R7	47K 1/4-watt 5% resistor
R3	10K 1/4-watt 5% resistor
R4	see text

❖ 3
Audio projects

SINCE OP AMPS ARE A FORM OF AMPLIFIER, THEY CAN BE USED IN many standard amplifier applications. Operational amplifiers tend to be very well suited to audio amplifier applications, because their frequency response is quite flat throughout the audio spectrum.

For serious applications, the 747 or 741 op amps may not be satisfactory. They will often add a noticeable amount of hiss or noise to the signal, which may be objectionable in some cases, but this would scarcely be a problem in hobbyist experiments.

In critical audio applications, you should substitute high-grade low-noise op amp devices for the 747 ICs specified here. The TL080 is generally a good choice. No other circuit changes will be required in most cases. To be on the safe side, check the manufacturer's specification sheet for the particular op amp device you intend to use in your circuit.

AUDIO PREAMP

The circuit shown in Fig. 3-1 is a preamplifier that can be used in many audio frequency applications. It can be used in radios, musical instrument amplifiers, or stereo systems (a complete pre-amplifier circuit will be needed for each channel). The high current main amplification would be performed by a separate power amplifier stage. A suitable parts list for this project is given in Table 3-1.

Overall circuit gain is set by potentiometer R6. Between the two op amp stages is a passive filter network that allows the user some control over the frequency response of the circuit. The high-frequency (treble) range can be cut back somewhat by adjusting potentiometer R7. Similarly, the potentiometer R11

50 Audio projects

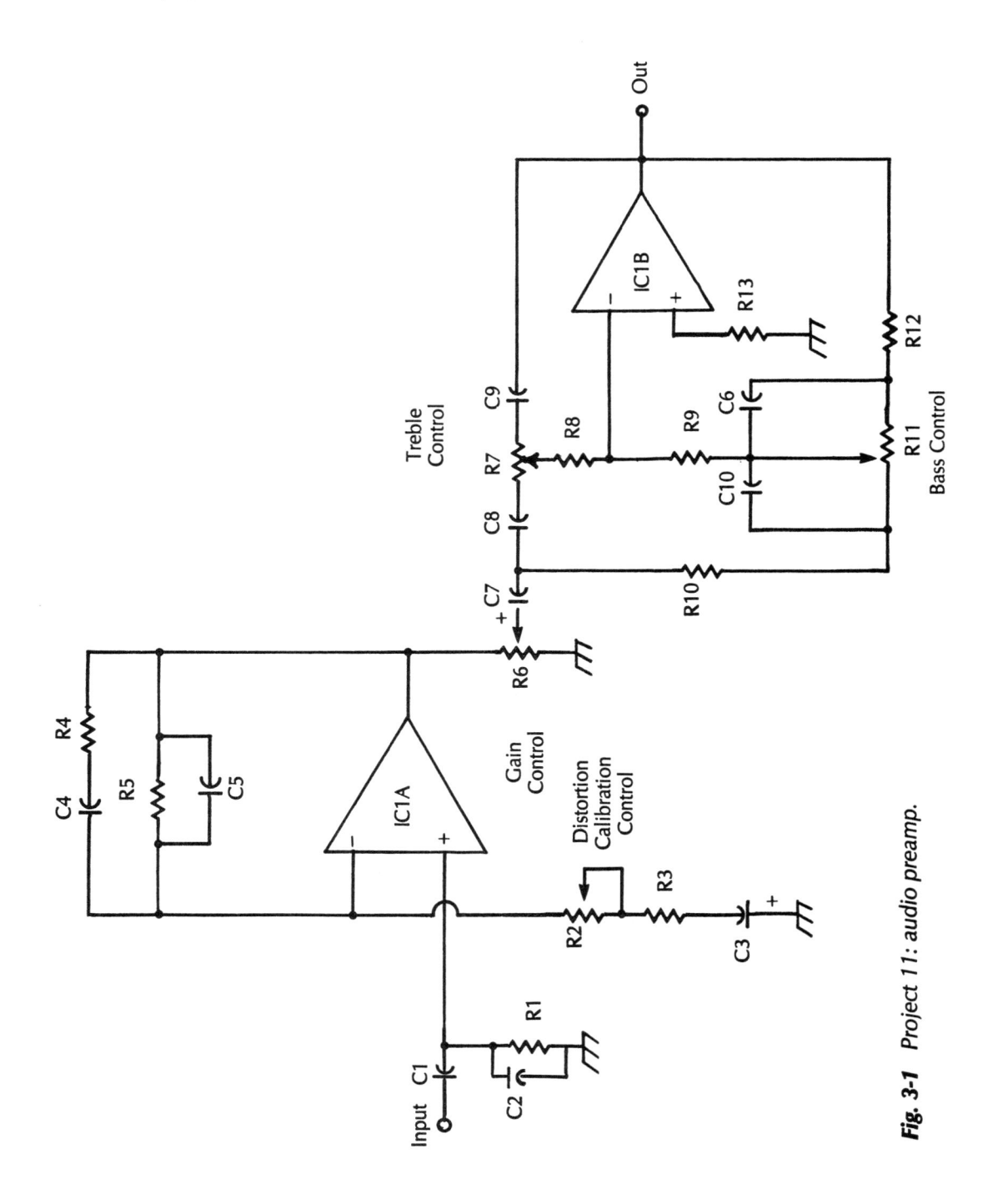

Fig. 3-1 *Project 11: audio preamp.*

Table 3-1. Parts List for
Project 11: Audio Preamp.

Part	Description
IC1	747 dual op amp
C1	1 µF 50-volt electrolytic capacitor
C2	50 pF capacitor
C3	68 µF 50-volt electrolytic capacitor
C4	0.1 µF capacitor
C5, C8, C9	0.003 µF capacitor
C6, C10	0.03 µF capacitor
C7	47 µF 50-volt electrolytic capacitor
R1	47K 5% 1/4-watt resistor
R2	100 ohm trimpot
R3	100 ohm 5% 1/4-watt resistor
R4	27K 5% 1/4-watt resistor
R5	220K 5% 1/4-watt resistor
R6	5K potentiometer
R7, R11	100K potentiometer
R8	3.3K 5% 1/4-watt resistor
R9, R10, R12	10K 5% 1/4-watt resistor
R13	68K 5% 1/4-watt resistor

trols the low-frequency (bass) response of the preamp. Trimpot R2 is adjusted during calibration for minimum output distortion. This should not be a front panel control. Once it has been set correctly, this trimpot should be left alone.

The 747 op amps will work in this circuit for a low-end application. For serious audio work, substitute op amps such as the TL080 for both IC1A and IC1B.

STEREO MAGNETIC CARTRIDGE PREAMP

Dual op amp ICs, such as the 747 can be particularly handy in stereophonic (two channel) audio applications. One of the op amps can be used in the right channel, and the other in the left channel. This minimizes the overall parts count and project size. Often, only a handful of external resistors and capacitors are required to complete the circuit.

Figure 3-2 shows the complete schematic diagram for a stereo magnetic cartridge preamplifier. Magnetic cartridges offer greater fidelity than ceramic cartridges. Since they use considerably less pressure, they also put less wear and tear on the record

52 Audio projects

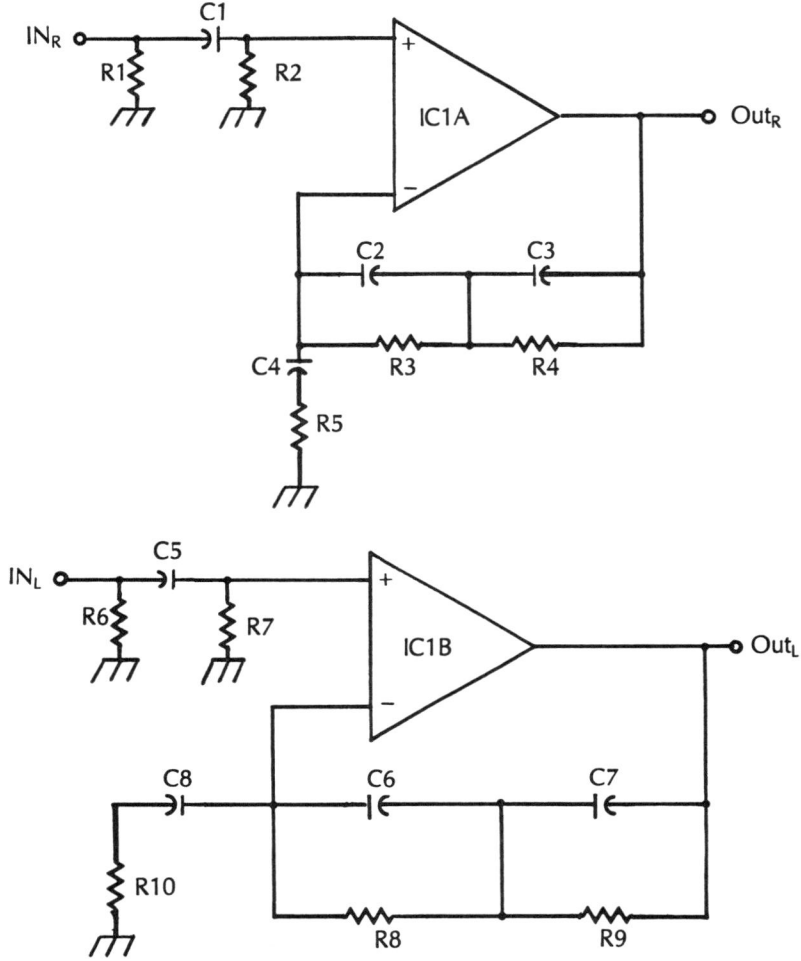

Fig. 3-2 Project 12: stereo magnetic cartridge preamp.

being played. The output signal from a magnetic cartridge is typically very low, compared to a ceramic cartridge, so an extra preamplification stage is generally required. A suitable parts list for this project is given in Table 3-2.

This circuit has two identical halves. IC1A, capacitors C1 through C4, and resistors R1 through R5 make up the right channel. The left channel is comprised of IC1B, capacitors C5 through C8, and resistors R6 through R10. Only the power supply is common to both halves of the circuit.

Table 3-2. Parts List for Project 12:
Stereo Magnetic Cartridge Preamp.

Part	Description
IC1	747 dual op amp
C1, C5	0.22 µF capacitor
C2, C6	0.0056 µF capacitor
C3, C7	0.0015 µF capacitor
C4, C8	22 µF 50-volt electrolytic capacitor
R1, R6	47K 1/4-watt 5% resistor
R2, R7	820K 1/4-watt 5% resistor
R3, R8	750K 1/4-watt 5% resistor
R4, R9	51K 1/4-watt 5% resistor
R5, R10	1K 1/4-watt 5% resistor

Using the component values recommended in the parts list, the stereo magnetic cartridge preamp will feature standard RIAA de-emphasis. The input signal is passed through a simple filter network made up of capacitor C1 (C5), and resistors R1 (R6) and R2 (R7). This filtering decreases the gain for very low frequencies in accordance with the RIAA standards. For maximum precision, replace resistors R2 and R7 with trimpots and adjust for a minimum offset at the output.

TAPE HEAD PREAMP

Another stereo preamplifier circuit is shown in Fig. 3-3. This project is designed for use with a magnetic tape head. Once again, low frequencies are given less gain than high frequencies. The de-emphasis/equalization curve in this case is according to NAB standards, used in most modern tape players. A high-pass filter is formed by capacitor C2 (C5) and resistor R3 (R8) in the feedback path of each op amp. This results in a boost of bass (low) frequencies, and a controlled cut of treble (high) frequencies.

The parts list for this project appears as Table 3-3. The circuit is made up of two identical halves. Op amp IC1A, capacitors C1 through C3, and resistors R1 through R5 make up the right channel, while the left channel is comprised of IC1B, capacitors C4 through C6, and resistors R6 through R10.

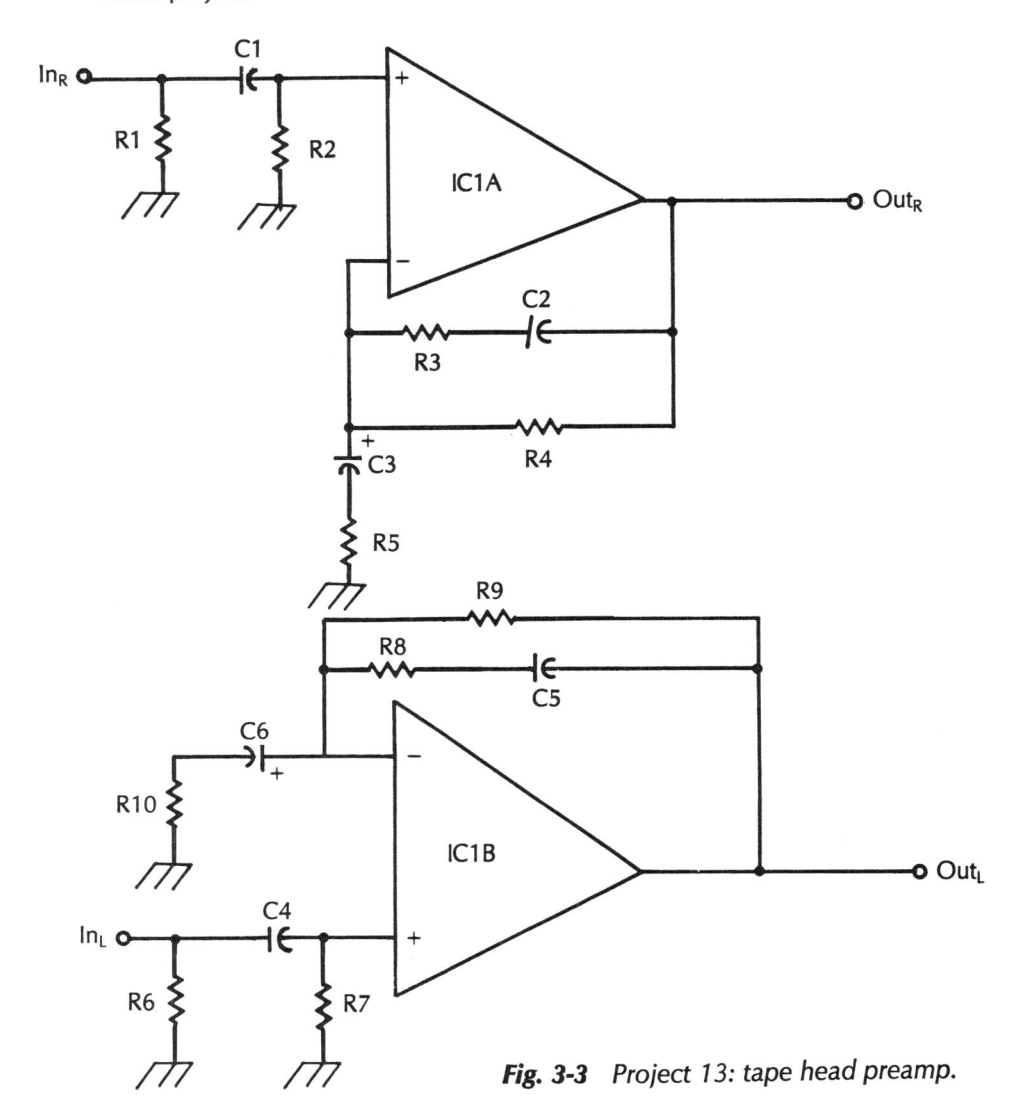

54　Audio projects

Fig. 3-3　Project 13: tape head preamp.

AUDIO TONE CONTROL

Tone controls in an audio system are a nice extra feature. They permit you to "customize" the sound, emphasizing or de-emphasizing bass (low) and treble (high) frequencies to your individual tastes, or the particular acoustic characteristics of the room. A stereo audio tone control circuit is shown in Fig. 3-4. A suitable parts list for this project is given in Table 3-4.

Table 3-3. Parts List for
Project 13: Tape Head Preamp.

Part	Description
IC1	747 dual op amp
C1, C4	0.22 µF capacitor
C2, C5	0.0047 µF capacitor
C3, C6	22 µF 50-volt electrolytic capacitor
R1, R6	47K 1/4-watt 5% resistor
R2, R7	1.5M 1/4-watt 5% resistor
R3, R8	51K 1/4-watt 5% resistor
R4, R9	820K 1/4-watt 5% resistor
R5, R10	1K 1/4-watt 5% resistor

This circuit consists of two identical halves, one for the right channel and one for the left channel. The tone adjustment is fully independent for each channel. Bass and treble boost and cut are individually controllable within each channel. Potentiometer R2 controls the right channel bass, while potentiometer R7 performs the same function in the left channel. The treble is controlled via potentiometer R4 in the right channel and R9 in the left channel.

AUDIO MIXER

In many audio applications, it is necessary to combine signals from two or more sources. This can be done with a mixer circuit,

Table 3-4. Parts List for
Project 14: Audio Tone Control.

Part	Description
IC1	747 dual op amp
C1, C2, C5, C6	0.05 µF capacitor
C3, C4, C7, C8	0.01 µF capacitor
R1, R3, R6, R8	10K 1/4-watt 5% resistor
R2, R7	100K potentiometer
R4, R9	50K potentiometer
R5, R10	1K 1/4-watt 5% resistor

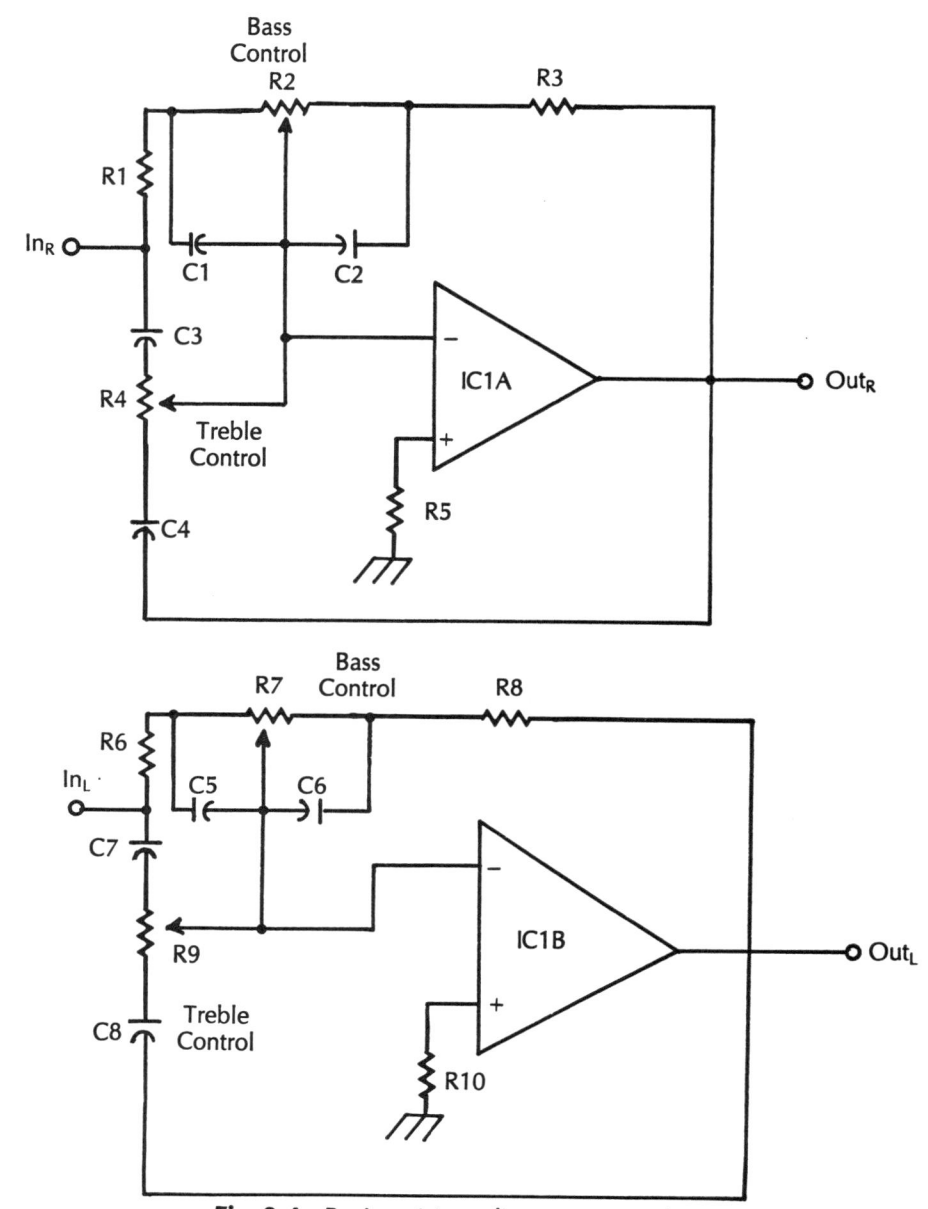

Fig. 3-4 *Project 14: audio tone control.*

like the one shown in Fig. 3-5. A suitable parts list for this project appears as Table 3-5.

This circuit is quite simple. IC1A is a simple summing amplifier. Four inputs are shown here, but you can easily adapt

Audio mixer

Fig. 3-5 Project 15: audio mixer.

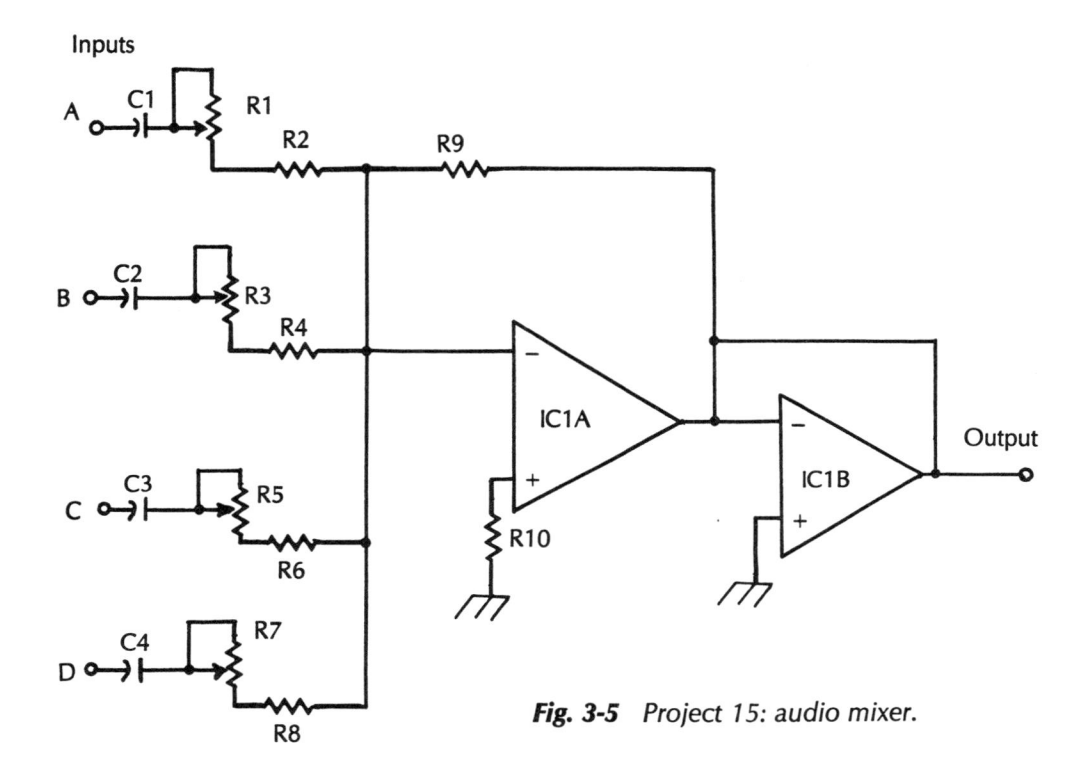

Table 3-5. Parts List for
Project 15: Audio Mixer.

Part	Description
IC1	747 dual op amp
C1, C2, C3, C4	0.22 µF capacitor
R1, R3, R5, R7	100K potentiometer (audio taper)
R2, R4, R6, R8	1K 1/4-watt 5% resistor
R9	10K potentiometer
R10	4.7K 1/4-watt 5% resistor

the circuit for a different number of inputs. Each input line includes a small capacitor (C1 through C4) to block any possible dc component in the signal, a fixed resistor (R2, R4, R6, and R8) and a potentiometer (R1, R3, R5, and R7). The potentiometers permit the user to independently adjust the levels of each of the input signals. A master level (or volume) control is potentiometer

R9 in the feedback path of IC1A. This control adjusts the overall output level. All of the input signals are affected equally by this master control.

The second stage (IC1B) is an inverting voltage follower to act as an output buffer, and adjust the output signal so that it is in phase with the original input signals. This circuit is not frequency dependent, and should exhibit a fairly flat frequency response across the entire audio range. In high-fidelity applications requiring full bass response, you might want to omit the input capacitors from the circuit.

❖ 4
Signal generator projects

SIGNAL GENERATORS ARE ALWAYS A POPULAR TYPE OF ELECTRONIC project. They are used in communications, recording, automated control systems, testing, electronic music, and many other applications. In fact, it is rare when electronic systems of any size and complexity don't have at least one signal generator stage.

Signal generators are sometimes called *oscillators,* although, strictly speaking, an oscillator produces only sine waves. A signal generator may produce any recurring ac waveform. The simplest ac waveform is the sine wave which consists of only a single (*fundamental*) frequency. Other ac waveforms are more complex, with additional frequency components, known as *harmonics*. A harmonic is some integer multiple of the related fundamental frequency.

If excessive positive feedback gets into an amplification loop, the amplifier circuit will break into oscillations. Although this is the simplest type of signal generator, it is undesirable because it shows up where such a generated signal is unwanted. The effect is usually more or less completely uncontrollable.

Signal generators can, depending upon their specific circuit design, produce ac signals in any frequency range, from the subaudible (below 20 Hz), to the audible region (from about 20 Hz to approximately 20 kHz), to the radio frequency (RF) range (up to several megahertz), and even up to the microwave region (hundreds of megahertz). The desired frequency range will depend upon the specific intended application. No single practical signal generator circuit has ever been designed to cover all possible frequency ranges. Different basic circuit designs are needed for different general ranges of frequencies. A distinction usually is made between AF (audio frequency) and RF (radio frequency) signal generator circuits. Sub-audio signal generators typically use the same basic designs as audio signal generators, only with larger capacitor and resistor values. Microwave circuits almost always require specialized components and circuit design techniques.

This chapter will feature a number of signal generator circuits of different types, producing a variety of useful ac waveforms. The signal generator projects presented in this chapter are designed primarily for AF (audio frequency) applications.

QUADRATURE OSCILLATOR

True oscillator circuits generate sine waves that are used in many applications, including test signals, control signals, carrier waves (see chapter 8), and additive synthesis. A specialized type of sine wave oscillator is the quadrature oscillator. This type of oscillator has two output signals, both at the same frequency. One of the outputs is a straight sine wave signal. The other output is a cosine wave. If you just look at a cosine wave by itself, it doesn't look any different from a sine wave. The only difference between the two signals is the phase. A cosine wave is phase shifted 90 degrees from the accompanying sine wave, as illustrated in Fig. 4-1. The phase shifted outputs can be used in complex control systems, and other applications.

A practical quadrature oscillator circuit built around a pair of op amps is shown in Fig. 4-2. A typical parts list for this project is given in Table 4-1. As you can see, this is a very simple circuit. You might want to build this project even if you only need a single straight sine wave signal. If you are using a dual op amp device, such as the 747, the parts count for a quadrature oscillator is typically less than a typical straight sine wave oscillator circuit of comparable signal quality. Aside from the two op amps, only six additional passive components are required in this circuit: three capacitors and three resistors.

Designing the circuit and choosing the component values is pretty straightforward, since all three capacitors have identical values in this circuit. Resistors R2 and R3 also have equal values. The value of resistor R1 should be slightly less than that of resistor R2 to ensure that oscillations will begin as soon as power is

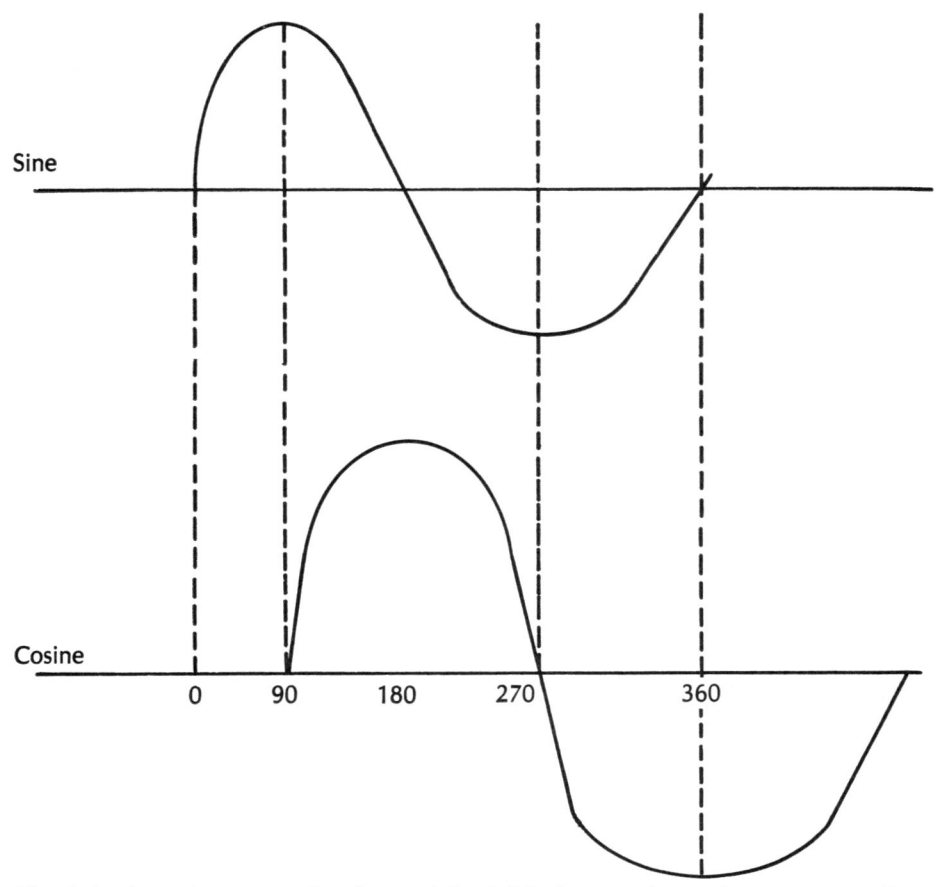

Fig. 4-1 *A cosine wave is phase shifted 90 degrees from the corresponding sine wave.*

applied to the circuit. The exact value of resistor R1 is not at all critical, as long as it is slightly less than the value of resistor R2.

This means that you only need to select two component values to design a working version of this quadrature oscillator circuit. One is the value for the capacitors C1, C2, and C3; and the other is the value for resistors R2 and R3. The formula for the output frequency generated by this oscillator circuit is fairly simple:

$$F = \frac{1}{2\pi RC}$$

Pi (π) is a mathematical constant with a value of approximately

Fig. 4-2 Project 16: quadrature oscillator.

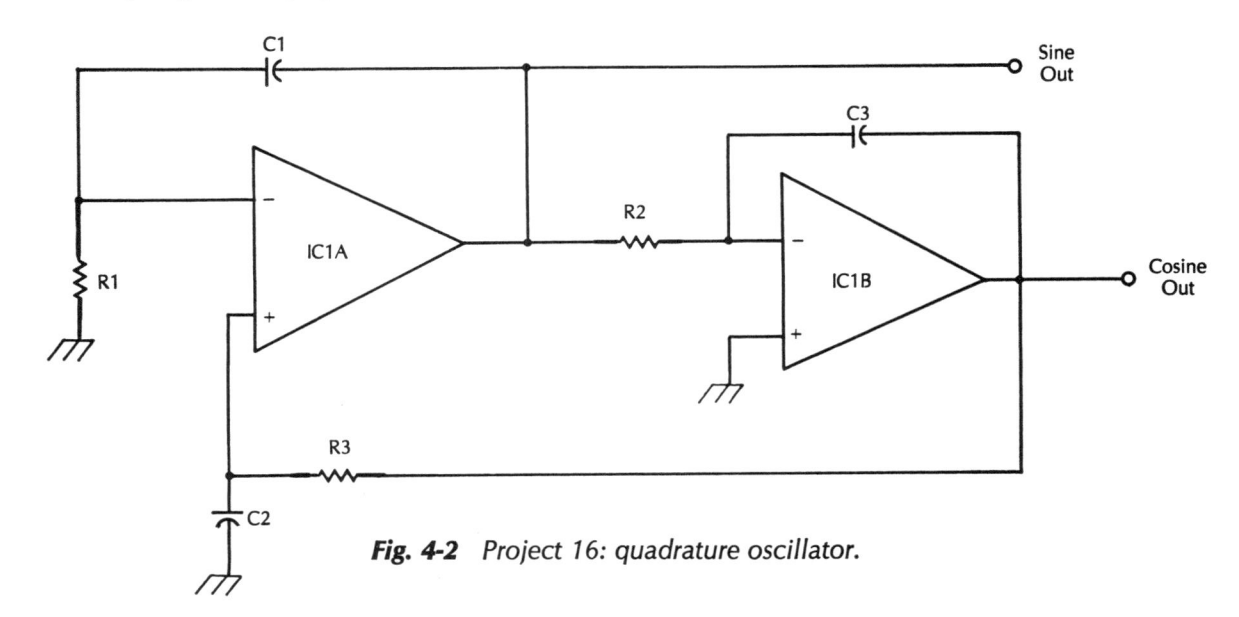

Table 4-1. Parts List for Project 16: Quadrature Oscillator.

Part	Description
IC1	747 dual op amp
C1 – C3	0.05 µF capacitor
R1	2.7K 5% ¼-watt resistor
R2, R3	3.3K 5% ¼-watt resistor

mately 3.14. The frequency equation for this circuit can therefore be rewritten as:

$$F = \frac{1}{6.28RC}$$

Using the component values suggested in the parts list, the circuit's output frequency works out to:

$$F = \frac{1}{6.28 \times 3300 \times 0.0000005}$$
$$= \frac{1}{0.0010362}$$
$$= 965 \text{ Hz}$$

You can round this off to about 1000 Hz (1 kHz). Component tolerances could account for as much error. Experiment with alternate component values to generate signals at different frequencies.

This quadrature oscillator circuit can easily be adapted to permit a manually adjustable output frequency. Simply replace resistors R2 and R3 with a dual potentiometer. The shafts of a dual potentiometer are ganged together, so both resistances will always be equal. Fixed resistors should be used in series with the potentiometer sections so that the frequency determining value of R2 can never drop below the fixed value of resistor R1. For best results, you should use small valued potentiometers for a relatively narrow range.

In a practical design situation, you should already know the desired signal frequency to be generated, and will need to determine the desired component values. The easiest method is to arbitrarily select a likely capacitor value (C), and then rearrange the frequency equation to solve for R:

$$R = \frac{1}{6.28FC}$$

This quadrature oscillator circuit features simplicity and elegance of design. The output signals are relatively pure. The distortion in the output waveforms will be negligible in all but the most critical applications. The sine wave and cosine wave outputs from this circuit will always be at exactly the same frequency. The only difference between the two output signals is their relative phase.

TRIANGLE WAVE GENERATOR

A circuit which generates a waveform other than a sine wave is usually, though not always, called a signal generator, rather than an oscillator. Another popular waveform used in a great many applications, is the *triangle wave,* which is shown in Fig. 4-3. A sine wave consists of the fundamental frequency component only. There is no harmonic content in a pure sine wave. A triangle wave contains the fundamental frequency, and all the odd harmonics, at relatively weak amplitudes.

A triangle wave can be used in many of the same applications as a sine wave. When converted into accoustic energy

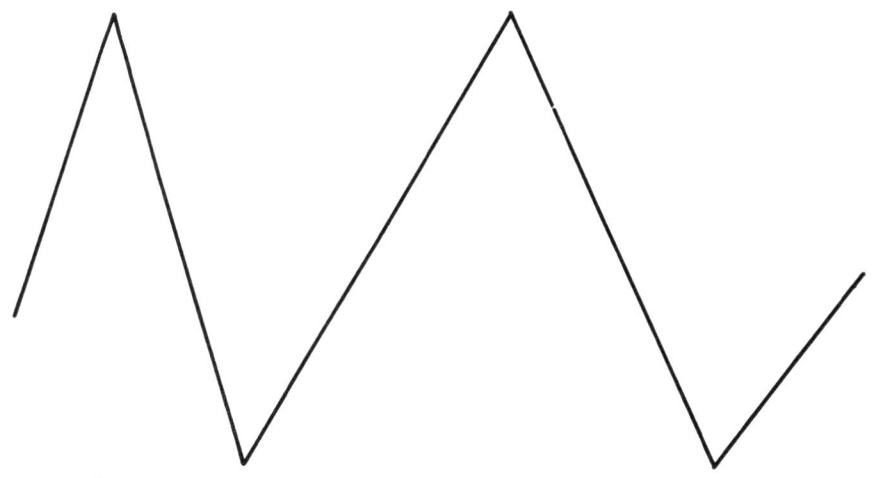

Fig. 4-3 *A triangle wave is a useful signal in many applications.*

through a loudspeaker, a triangle wave is much less wearing on the ear than a sine wave.

A *square wave,* shown in Fig. 4-4, has the same basic harmonic make up as a triangle wave (fundamental and all odd harmonics), except the harmonics have much stronger amplitudes in the square wave than in the triangle wave.

Square waves are very easy to generate with a single op amp. A pair of op amps can be used to generate triangle waves. Since a triangle wave is so similar to a square wave, it is not difficult at all to convert one into the other. Passing a square wave signal through a simple low-pass filter (or integrator) will produce a triangle wave signal at the output, by weakening the harmonics.

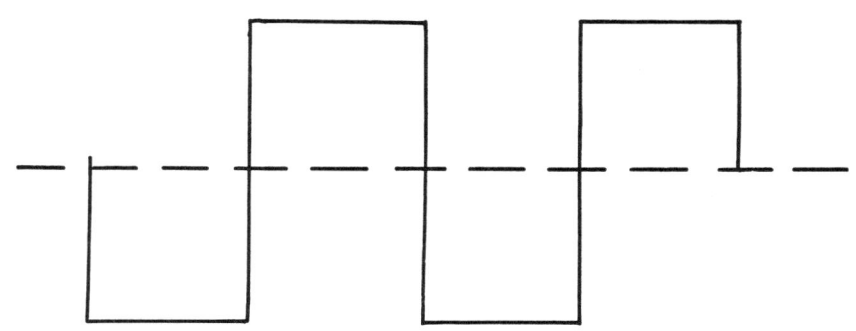

Fig. 4-4 *A square wave has a similar harmonic makeup to a triangle wave, but the harmonics are stronger.*

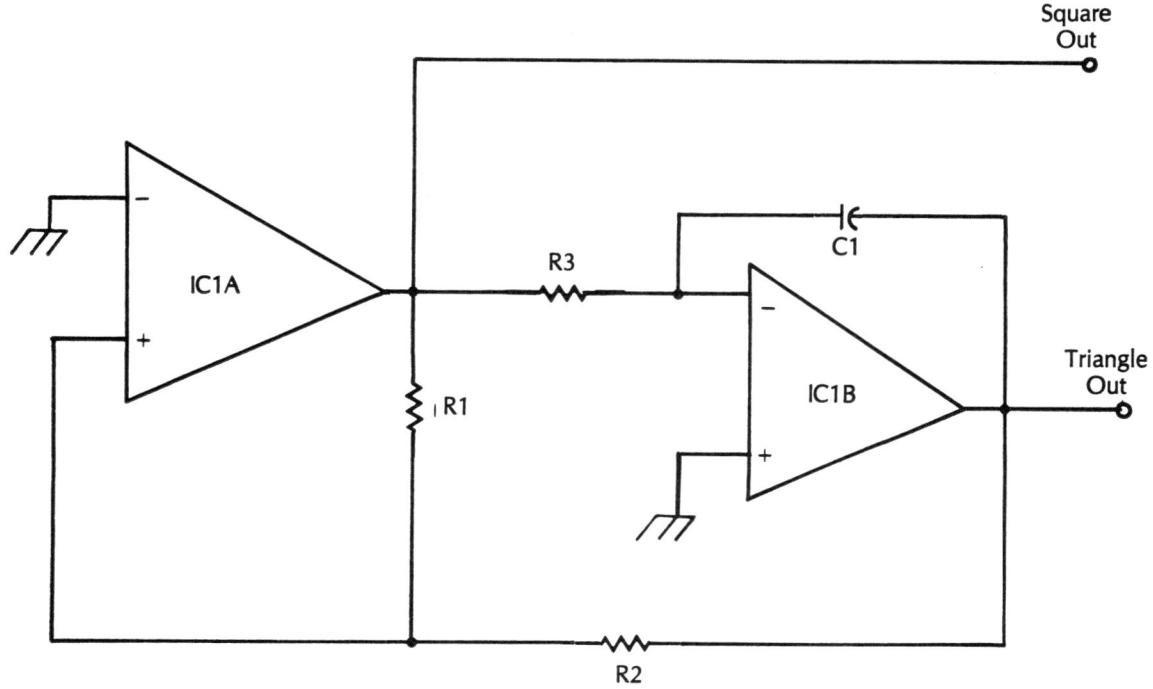

Fig. 4-5 *Project 17: triangle wave generator.*

**Table 4-2.
Parts List for Project 17:
Triangle Wave Generator.**

Part	Description
IC1	747 dual op amp
C1	0.1 µF capacitor
R1	22K 5% 1/4-watt resistor
R2	10K 5% 1/4-watt resistor
R3	2.7K 5% 1/4-watt resistor

Most op amp based triangle wave generator circuits use this basic approach.

A practical two op amp triangle wave generator circuit is shown in Fig. 4-5. A suitable parts list for this project is given in Table 4-2. As always, you can experiment with alternate component values in this circuit.

In this circuit, op amp IC1A is wired as a comparator. It generates a square wave signal, when the circuit's output is fed back into the noninverting input of this op amp. This square wave signal is then passed through op amp IC1B, which is wired as an integrator, or low-pass filter. The filter decreases the amplitude of the signal's harmonics, converting the square wave into a triangle wave. Both waveforms (the square wave and the triangle wave) can be tapped off from the circuit simultaneously, and both output signals will be at the exact same frequency.

The square wave signal will swing back and forth between the op amp's positive and negative saturation voltages. The amplitude of the triangle wave signal is controlled by the ratio of resistors R2/R1. These components also help determine the signal frequency, so making any adjustments in this circuit can be a little tricky. The formula for finding the output frequency (for both waveform signals) is:

$$F = \frac{1}{4R_3C_1} \times \frac{R_1}{R_2}$$

Using the suggested component values from the parts list, we get an output frequency of:

$$F = \frac{1}{4 \times 2700 \times 0.0000001} \times \frac{22{,}000}{10{,}000}$$
$$= \frac{1}{0.00108} \times 2.2$$
$$= 926 \times 2.2$$
$$= 2037 \text{ Hz}$$

A signal generator circuit that can put out more than one of the basic waveforms is known as a *function generator*. This project qualifies for the title by this definition.

TWO-FREQUENCY OSCILLATOR

Figure 4-6 shows a very unusual signal generator circuit. This circuit has two separate, and independently usable outputs. The frequency of the signal at output A is twice the frequency of the signal at output B. That is:

$$A = 2 \times B$$

or

$$B = A/2$$

Fig. 4-6 *Project 18: two-frequency oscillator.*

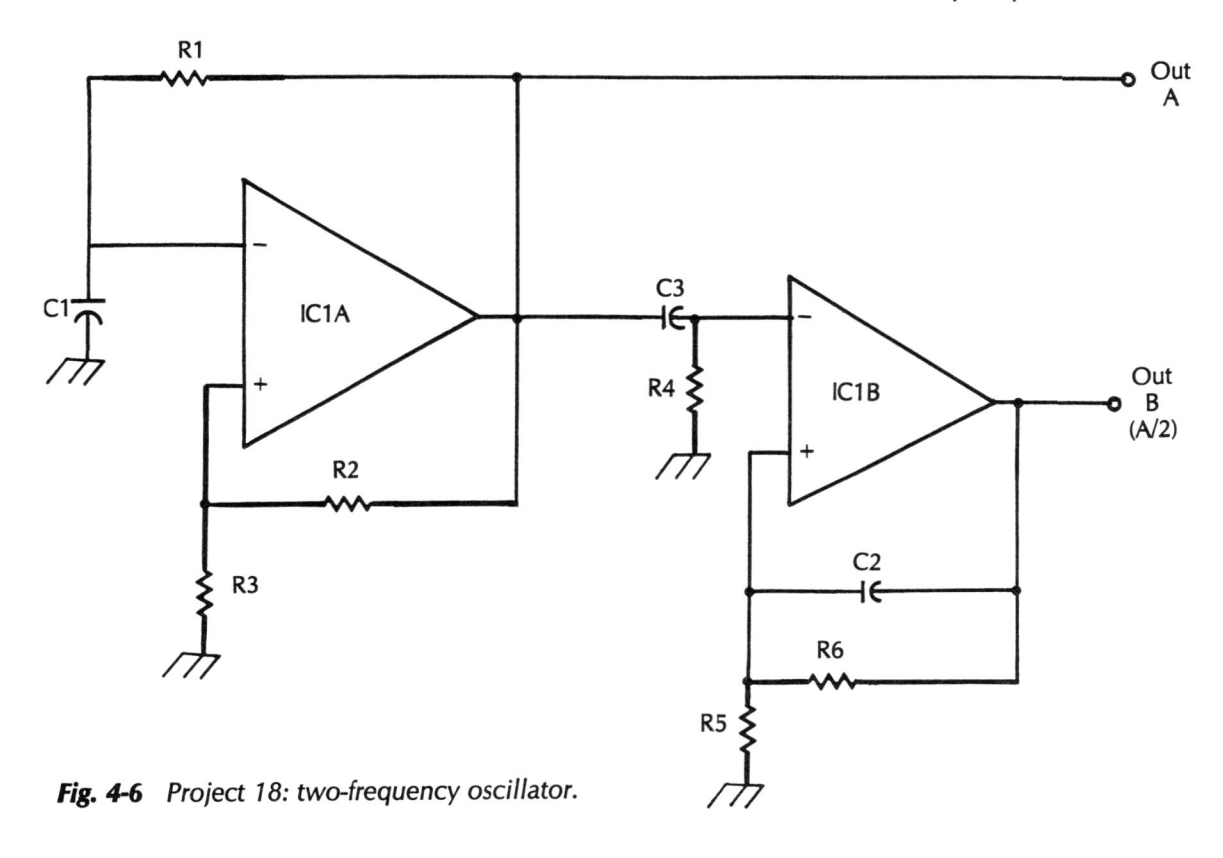

Table 4-3. Parts List for Project 18: Two-Frequency Oscillator.

Part	Description
IC1	747 dual op amp
C1, C2, C3	0.1 µF capacitor
R1	47K 5% 1/4-watt resistor
R2, R5	100K 5% 1/4-watt resistor
R3	10K 5% 1/4-watt resistor
R4	1K 5% 1/4-watt resistor
R6	1M 5% 1/4-watt resistor

Both signals are in the form of square waves.

A suitable parts list for this unusual project appears as Table 4-3. Basically, IC1A and its associated components (capacitor C1, and resistors R1, R2, and R3) form a simple square wave generator. Assuming that resistor R2 has a value equal to ten times the value of resistor R3, the generated frequency will be equal to:

$$F_a = \frac{5}{R_1 C_1}$$

Changing the ratio of values for resistors R2 and R3 will alter the frequency equation. It is best to hold these values constant during your experimentation with the circuit. There is no sense in overcomplicating things unnecessarily.

Using the component values suggested in the parts list, the signal frequency generated by this stage of the circuit should be equal to:

$$F_a = \frac{5}{47000 \times 0.0000001}$$
$$= \frac{5}{0.00047}$$
$$= 1064 \text{ Hz}$$

For most practical purposes, this can be rounded off to about 1000 Hz (1 kHz).

This 1 kHz signal is available directly at output A. This signal is also fed into the second stage of the circuit, comprised of IC1B, and its associated components (capacitors C2 and C3, and resistors R4, R5, and R6).

This stage is a bistable multivibrator or *flip-flop*. Each time the input of a flip-flop is triggered, its output reverses state. If the flip-flop's output starts out low, it will go high on input pulse 1, low on input pulse 2, high again on input pulse 3, and so forth. The process is illustrated in Fig. 4-7.

As you can see, for every two input pulses, there is one output pulse. In other words, if a constant stream of regular pulses is fed into the flip-flop's input, the output will be a constant stream of regular pulses at exactly one-half the original frequency. This half-frequency signal is tapped off from the circuit at output B. Using the component values from the suggested parts list, the signal frequency at output B should be equal to:

$$F_B = F_A/2$$
$$= 1064/2$$
$$= 532 \text{ Hz}$$

Again, we can round this off to a more convenient 500 Hz. Exper-

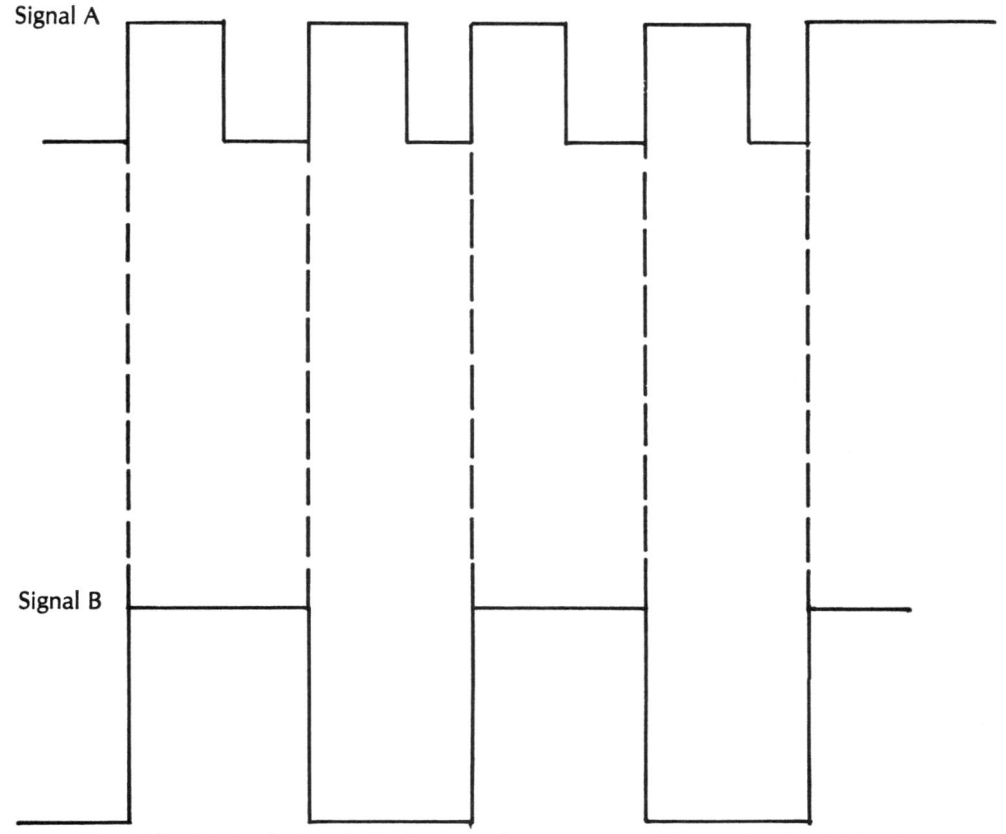

Fig. 4-7 *Typical signals in the two-frequency oscillator circuit of Fig. 4-6.*

iment with alternate values for capacitor C1 and resistor R1 to generate other signal frequencies with this circuit.

SAWTOOTH WAVE GENERATOR

Another common ac waveform useful in many applications is the *sawtooth wave*, or *ramp wave*, illustrated in Fig. 4-8. The signal voltage starts at a minimum value, builds steadily and linearly up to a maximum value, then quickly drops back down to the minimum value again to start the next cycle.

Actually, there are two types of sawtooth waves. The one we have looked at here is known as the *ascending* sawtooth wave. A *descending* sawtooth wave works in basically the same way, but backwards. The signal voltage in this waveform starts at the maximum value, drops steadily and linearly down to the minimum

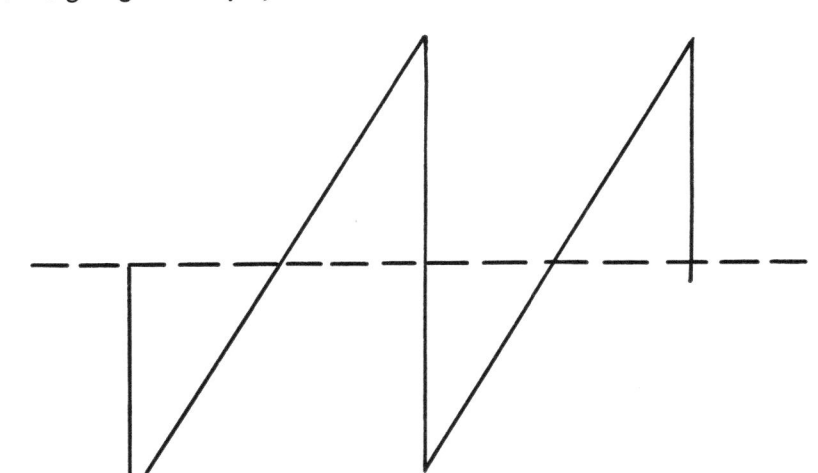

Fig. 4-8 *An ascending sawtooth wave is very rich in harmonics.*

value, then quickly jumps back up to the maximum value again to start the next cycle. A typical descending sawtooth wave is illustrated in Fig. 4-9.

Sawtooth waves are very rich in harmonics. All harmonics (both even and odd) are included in this waveform, and start at a fairly strong level. The higher a harmonic is, the weaker its amplitude will be.

Sawtooth waves have a great many practical applications. One of the most familiar is to move an electron beam across the

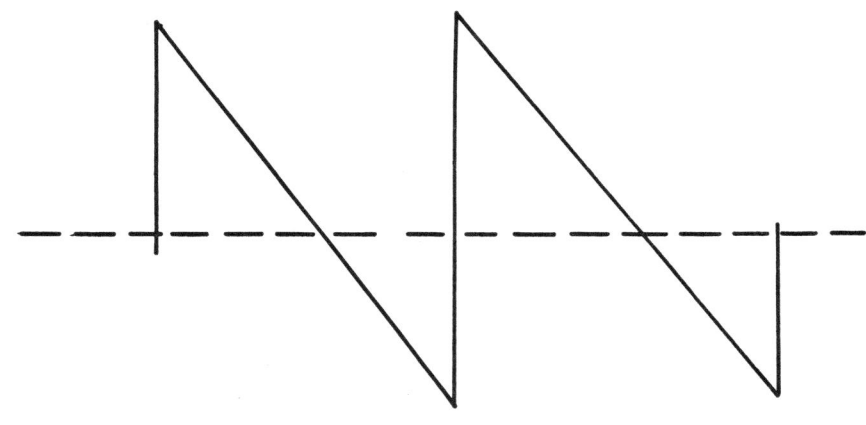

Fig. 4-9 *A descending sawtooth wave is also possible.*

screen of a television set or an oscilloscope. This waveform is also used in certain testing procedures. Other applications for sawtooth waves include automation control and electronic music.

A complete sawtooth wave generator circuit is shown in Fig. 4-10. A suitable parts list for this project is given in Table 4-4.

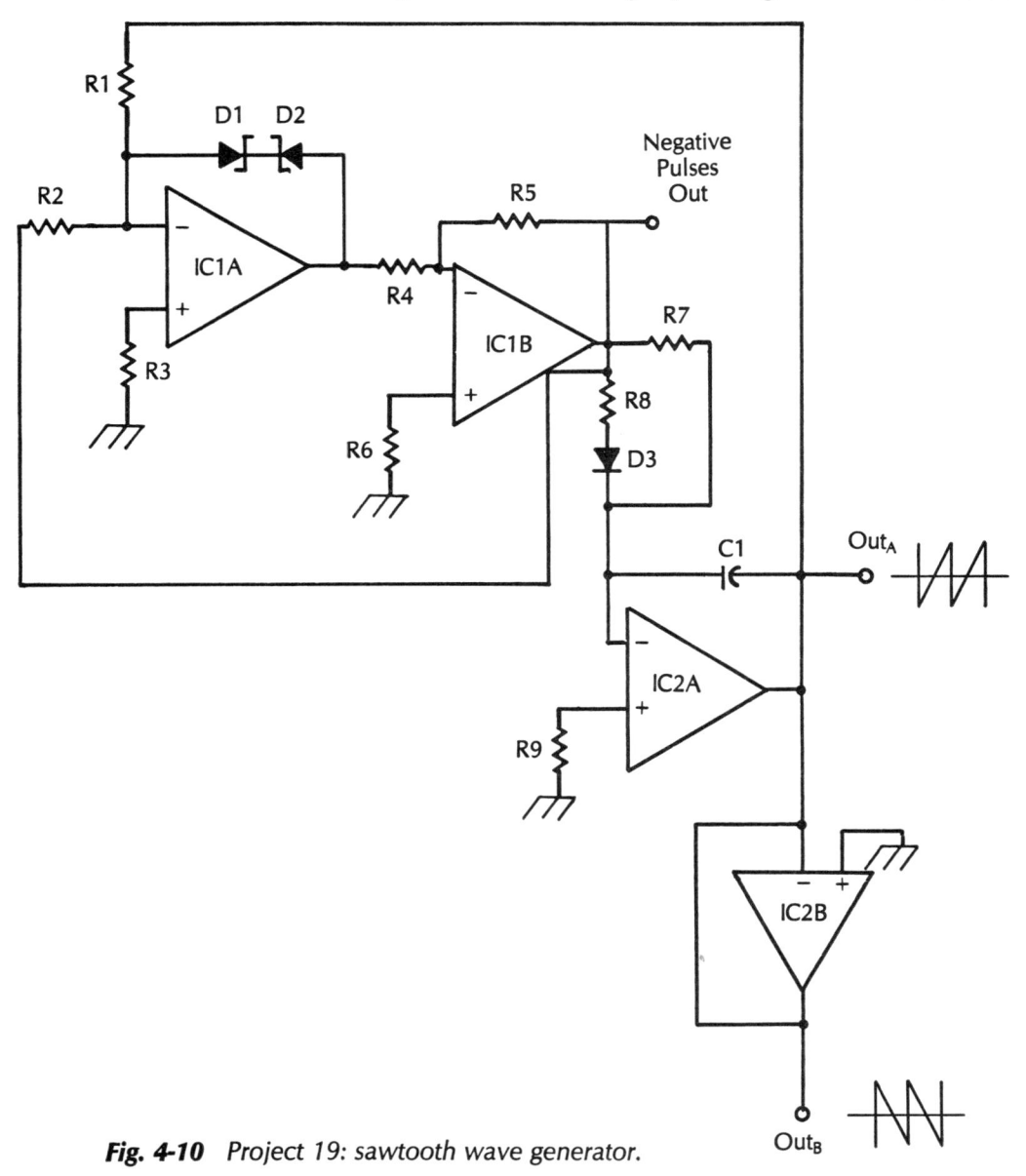

Fig. 4-10 Project 19: sawtooth wave generator.

Signal generator projects

**Table 4-4.
Parts List for Project 19:
Sawtooth Wave Generator.**

Part	Description
IC1, IC2	747 dual op amp
D1, D2	zener diode (5.1 volt)
D3	diode (1N4148, or similar)
C1	0.1 µF capacitor
R1, R2	10K 1/4-watt 5% resistor
R3, R6	4.7K 1/4-watt 5% resistor
R4, R5	10K 1/4-watt 5% resistor
R7, R9	18K 1/4-watt 5% resistor
R8	1K 1/4-watt 5% resistor

You will probably want to experiment with alternate component values in this project.

This sawtooth wave generator circuit features two outputs. Output A puts out an ascending sawtooth wave, and output B puts out a descending sawtooth wave. Both signals are at the exact same frequency. The descending sawtooth wave at output B is derived simply by inverting the ascending sawtooth wave.

IC1A is a linear integrator. IC1B and IC2A form a feedback loop to cause different integration rates for the ascending and descending portions of the cycle. Resistor R8, which is functionally part of the circuit only when diode D3 is forward-biased, has a very small value (compared to resistor R7), so the descending portion of the cycle is much faster than the ascending portion. It won't be a truly instantaneous drop from the maximum level back down to the minimum level, but it will be very fast, and should be good enough for most practical applications.

The approximate signal frequency can be found with this formula:

$$F = \frac{1}{R_7 C_1}$$

Using the component values suggested in the parts list, the circuit will put out a sawtooth wave signal with a frequency of about:

$$F = \frac{1}{18,000 \times 0.0000001}$$

$$= 1/0.0018$$
$$= 550 \text{ Hz}$$

Zener diodes D1 and D2 set the minimum and maximum voltages of the output signal. Using 5.1-volt zener diodes, as specified in the parts list will result in about a 10-volt peak-to-peak output signal, ranging from −5 volts to +5 volts.

The main output signal from the circuit is tapped off at output A. This is an ascending sawtooth wave. This signal is also fed through an inverting voltage follower (IC2B) to produce a descending sawtooth wave at output B. If the two output signals (A and B) are mixed equally, the net result will be a continuous 0 volts. The two signals will always exactly cancel one another out. Yet another signal output may be tapped off from the output of IC1B. The signal at this point will be a string of narrow positive going pulses.

❖ 5
Filter projects

FOR MOST ELECTRONIC AMPLIFIER APPLICATIONS, THE FREQUENCY response should be as flat as possible. That is, the circuit gain should not fluctuate with any changes in the frequency of the input signal. The amplitude is continuous from 0 Hz (dc), on up into the Megahertz region. Modern circuits can come astonishingly close to this ideal goal, but no circuit has a 100% flat frequency response. Some frequency components will tend to be emphasized over other frequency components, filters take advantage of this fact.

A *filter* is an electronic amplifier circuit which is designed deliberately for a specific nonflat frequency response pattern. Certain frequency components are passed by the filter circuit, other frequency components are blocked by the circuitry. An active filter circuit incorporates an amplifier stage of some kind. Typically, passed frequency components are boosted (or amplified), while blocked frequency components are attenuated, or sharply reduced in amplitude.

There are four basic standard filter patterns. Most (but not all) practical filter circuits operate according to one of these patterns:

- Low-Pass
- High-Pass
- Bandpass
- Band-Reject

Each of these names is self explanatory. A *low-pass filter* blocks high frequencies, but permits low-frequency components to pass through to the output. A *high-pass filter* operates in just the opposite manner—high-frequency components are passed, and low frequencies are blocked by the circuit. A *bandpass filter* passes only those frequencies within a specific band, or *range*. Any frequency component outside the specified band is blocked by the circuit.

The opposite of a bandpass filter is the *band-reject filter.* Any frequency component lying within the specified band or range is blocked by this circuit. If a frequency component falls outside this band, in either direction (high or low), it is passed on through to the circuit's output. Band-reject filters are sometimes called *notch filters*. In the projects of this chapter, you will be working with examples of each of the four basic filter circuit types, along with a couple of specialized filter circuits.

In an ideal filter circuit, all frequency components are either blocked fully, or passed fully. Passed frequencies all have full amplitude, while all blocked frequencies are attenuated completely as illustrated in Fig. 5-1. Real filter circuits don't quite live up to this ideal. They exhibit a finite cutoff slope. Frequency components near the nominal cutoff frequency are attenuated partially, at a specific rate. The steeper the cutoff slope, the better the filtering action.

Filter circuits are rated according to their *order*, which is a standardized measure of the cutoff slope. The higher the order

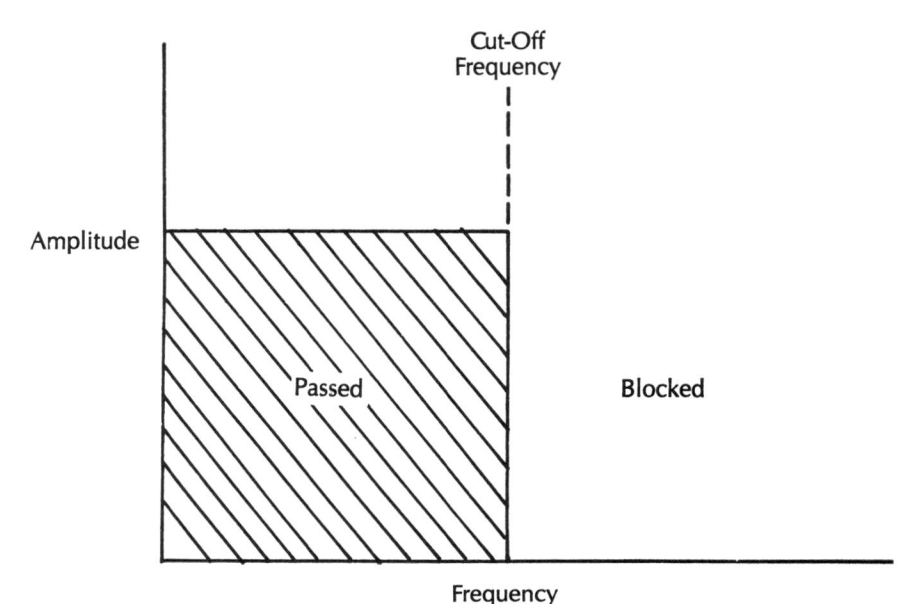

Fig. 5-1 An ideal low-pass filter has an infinite cutoff slope.

number, the steeper the cutoff slope, and the greater the attenuation of intermediate frequency components. The steeper the slope, the better the filter circuit.

The slope is defined as decibels of attenuation per octave (doubling of frequency). Active filter circuits commonly are divided into orders. Each order represents an attenuation of -6 dB per octave:

1st order	− 6 dB/octave
2nd order	− 12 dB/octave
3rd order	− 18 dB/octave
4th order	− 24 dB/octave
5th order	− 30 dB/octave

Higher order filters are much better at discriminating between passed and blocked frequencies, but the circuits tend to be more expensive and complex.

THIRD-ORDER LOW-PASS FILTER

For a low-pass filter, anything above the cutoff frequency is blocked, while anything below the cutoff frequency is passed. Frequencies components near the cutoff frequency are partially blocked, according to an attenuation, or cutoff slope. A frequency response graph for a practical low-pass filter is shown in Fig. 5-2.

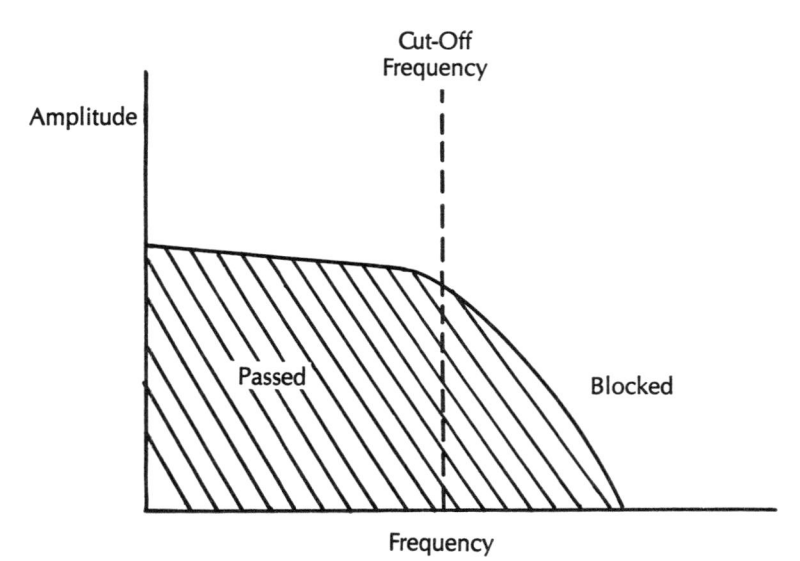

Fig. 5-2 *A practical low-pass filter circuit has a more gradual cutoff slope.*

Low-order filter circuits can be constructed around single op amps. A first-order low-pass filter circuit is shown in Fig. 5-3, and a second-order low-pass filter circuit is illustrated in Fig. 5-4.

Higher-order filters can be obtained by connecting low-order filters in series. This is where a dual op amp device, like the 747 can come in handy. A third-order filter can be created by placing

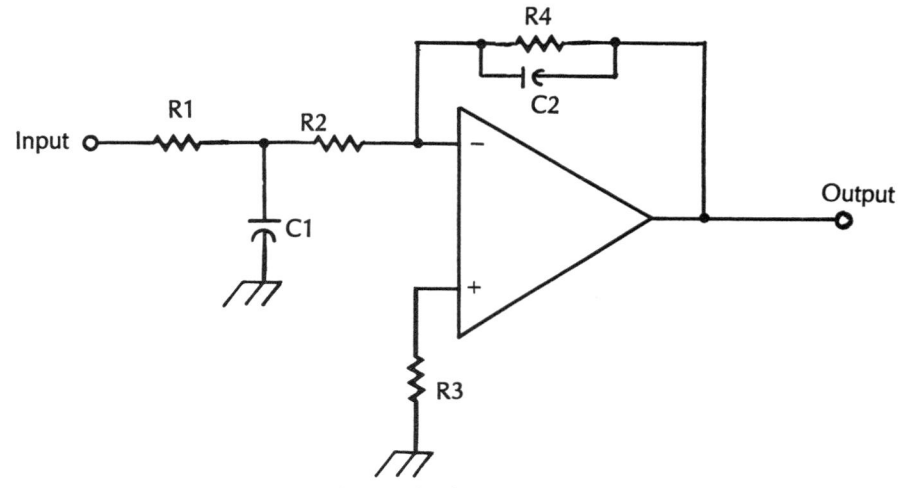

Fig. 5-3 A first-order low-pass filter circuit.

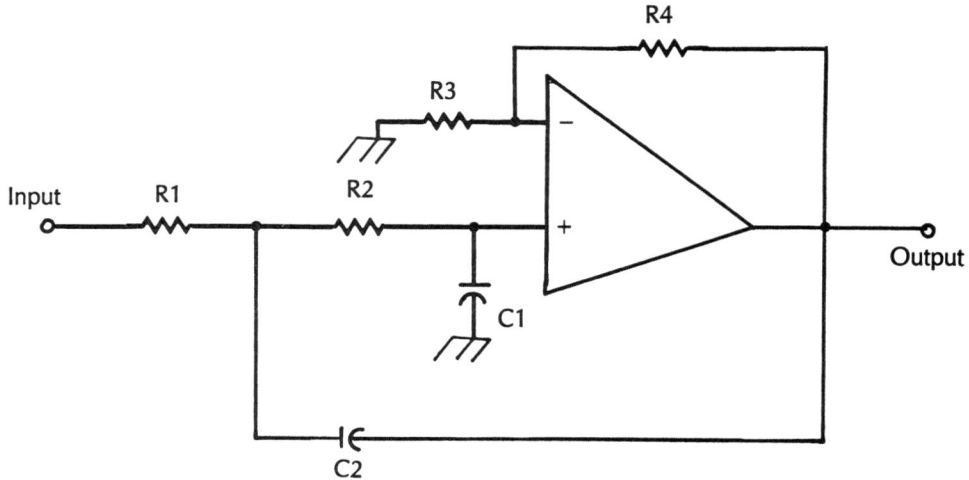

Fig. 5-4 A second-order low-pass filter circuit.

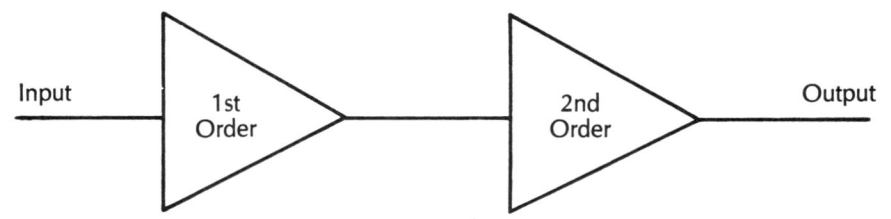

Fig. 5-5 *A third-order filter created by placing a first-order filter circuit in series with a second-order filter circuit.*

a first-order filter in series with a second-order filter, as shown in Fig. 5-5. Both component filters should be set for the same cutoff frequency for the circuit to function properly as a third-order filter.

The complete circuit for a low-pass, third-order filter appears in Fig. 5-6. A typical parts list for this project is given in Table 5-1. You should experiment with different values for the passive components in this circuit. The passive component values determine the cutoff frequency of the filter.

Compare the circuit of Fig. 5-6 to the circuits of Fig. 5-3 and Fig. 5-4. Notice that the third-order filter circuit is a first-order filter (IC1A and its associated components), followed by a second-order filter (IC1B and its associated components).

The first-order filter section is built around IC1A, along with six passive components. Those components are four resistors (R1

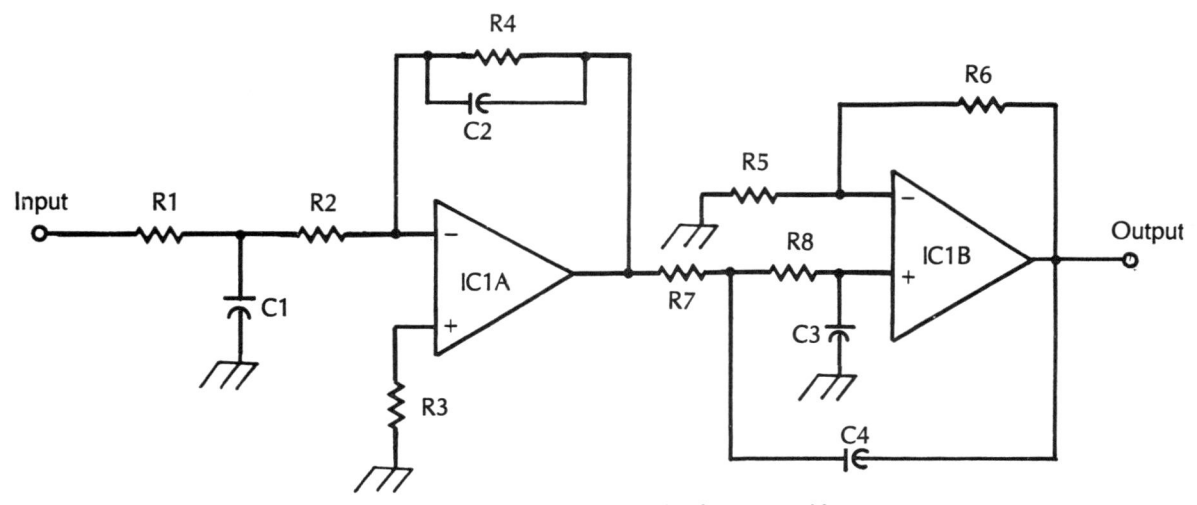

Fig. 5-6 *Project 20: third-order low-pass filter.*

Table 5-1. Parts List for Project 20: Third-Order Low-Pass Filter.

Part	Description
IC1	747 dual op amp
C1 – C4	0.1 µF capacitor
R1, R2, R4, R7, R8	1.5K 5% 1/4-watt resistor
R3	3.3K 5% 1/4-watt resistor
R5	1.8K 5% 1/4-watt resistor
R6	1K 5% 1/4-watt resistor

through R4), and two capacitors (C1 and C2). The values of these components interact. In order for the circuit to function properly, the following conditions must be true:

$$R_1 = R_2 = R_4$$
$$R_3 = 2R_1$$
$$C_1 = C_2$$

There is some leeway for component tolerances. High-precision resistors and capacitors aren't required for most applications. For example, in the parts list R1, R2, and R4 have a value of 1.5K. Nominally, resistor R3 should have a value of 3K, but this is not a standard resistor value, so a 3.3K resistor was used instead. It will throw off the cutoff frequency and the cutoff slope only slightly. The difference should be negligible, and the odds are you won't even be able to detect it.

The cutoff frequency for this portion of the circuit is determined by this formula:

$$F = \frac{1}{2\pi R_1 C_1}$$

Pi (π) has a value of approximately 3.14 so the cutoff frequency formula may be written as:

$$F = \frac{1}{6.28 R_1 C_1}$$

Using the suggested component values from the parts list, this gives a nominal cutoff frequency of:

$$F = \frac{1}{6.28 \times 1500 \times 0.0000001}$$

$$= \frac{1}{0.000942}$$
$$= 1062 \text{ Hz}$$

For convenience, round this off to 1000 Hz (1 kHz).

The second stage of this filter circuit is made up of IC1B, resistors R5 through R8, and capacitors C3 and C4. This stage is a second-order filter circuit. This stage requires the following component equalities:

$$R_6 = 0.586R_5$$
$$R_7 = R_8$$
$$C_3 = C_4$$

In order for this circuit to function, the gain of this stage must be 1.586. (The technical reasons for this are complex and will not be addressed here.) Substituting a trimpot for resistor R6 might be advisable in precision applications. The cutoff frequency for this portion of the circuit is:

$$F = \frac{1}{2\pi R_7 C_3}$$

or

$$F = \frac{1}{6.28 R_7 C_3}$$

Remember, both sections of this filter circuit must have the same nominal cutoff frequency. Using the component values from the parts list, the second stage of this filter circuit has a nominal cutoff frequency of:

$$F = \frac{1}{6.28 \times 1500 \times 0.0000001}$$
$$= \frac{1}{0.000942}$$
$$= 1062 \text{ Hz}$$

Notice that this is the same result you got with the first stage. Once again, you can round this off to 1000 Hz (1 kHz).

Notice how similar the cutoff frequency equations are for the two stages of this circuit:

$$F_1 = \frac{1}{6.28 R_1 C_1}$$

$$F_2 = \frac{1}{6.28 R_7 C_3}$$

The easiest way to work with this multistage filter circuit is to use similar component values in both stages. That is:

$$R_1 = R_7$$
$$C_1 = C_3$$

This will ensure that both stages are operating at the same cutoff frequency (allowing for minor differences due to component tolerances).

Serious audio applications are usually considered moderately critical. For best results, it is advisable to use 5 percent tolerance resistors and high-grade Mylar or polystyrene capacitors in any audio filter circuits. Avoid inexpensive ceramic disc capacitors because they tend to have fairly wide tolerances.

FOURTH-ORDER HIGH-PASS FILTER

The operation of a high-pass filter is just the opposite of a low-pass filter. In a high-pass filter, upper frequency components are passed on through to the output, while low-frequency compo-

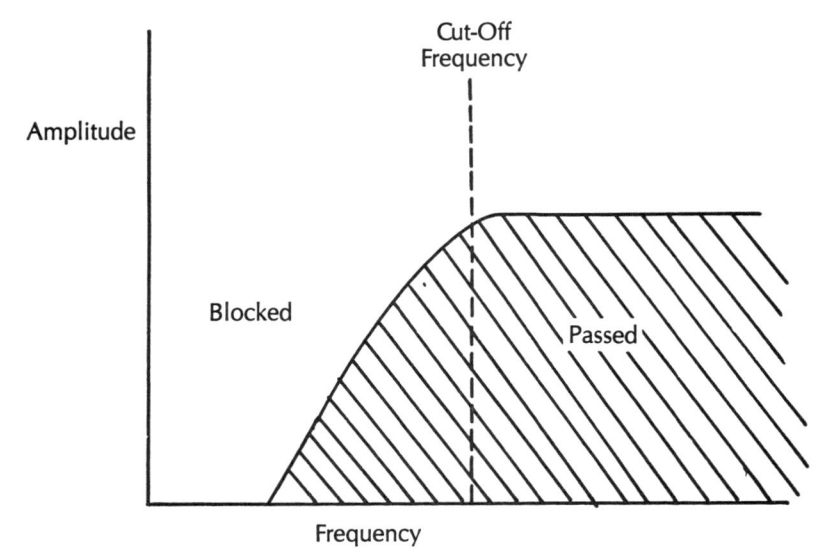

Fig. 5-7 *A high-pass filter functions in just the opposite manner as a low-pass filter.*

nents are blocked, or significantly attenuated. A frequency response graph for an ideal high-pass filter is illustrated in Fig. 5-7.

A practical high-pass filter circuit is shown in Fig. 5-8. A suitable parts list for this project is given in Table 5-2. Using the component values listed, the nominal cutoff frequency for the circuit will be about 1000 Hz (1 kHz). This filter circuit is a fourth-order type. This means that the cutoff slope falls off at a rate of 24 dB per octave.

For this circuit to work properly, resistors R1, R2, R5, and R6 should have identical values. Similarly, capacitors C1, C2, C3, and C4 should all have the same value. The nominal cutoff frequency for this high-pass filter circuit can be approximated using this formula:

$$F = \frac{1}{2\pi RC}$$

Pi (π), with a value of about 3.14, allows the cutoff frequency formula to be written as:

$$F = \frac{1}{6.28RC}$$

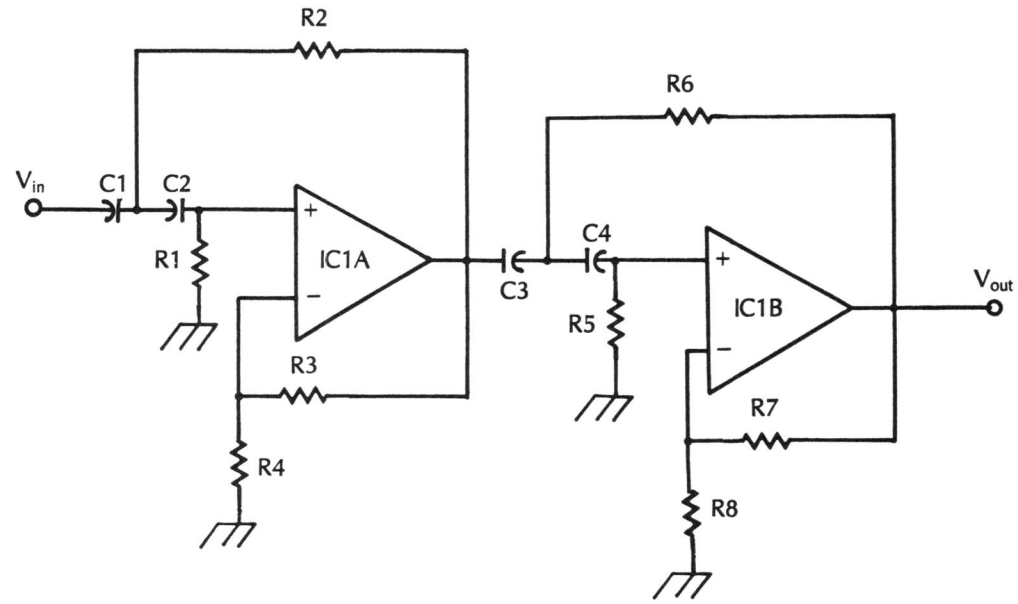

Fig. 5-8 *Project 21: fourth-order high-pass filter.*

Table 5-2. Parts List for Project 21: Fourth-Order High-Pass Filter.

Part	Description
IC1	747 dual op amp
C1, C2, C3, C4	0.005 µF capacitor
R1, R2, R5, R6	33K 1/4-watt 5% resistor
R3, R7	2.2K 1/4-watt 5% resistor
R4	15K 1/4-watt 5% resistor
R8	18K 1/4-watt 5% resistor

(nominal cutoff frequency = 1000 Hz)

In this formula, the resistance R should be in ohms and the capacitance C should be in farads. The parts list suggests the following component values:

$R = 33K$ (33,000 ohms)
$C = 0.005 \mu F$ (0.00000005 farad)

Using these recommended component values, the filter's nominal cutoff frequency works out to:

$$F = \frac{1}{6.28 \times 33000 \times 0.00000005}$$
$$= \frac{1}{0.0010362}$$
$$= 965 \text{ Hz}$$

For most applications, you can round this off to 1000 Hz (1 kHz). With the cutoff slope attenuating the signal amplitude as the frequency decreases, you can usually get away with a little fudging of the nominal cutoff frequency, except in very critical applications.

Resistors R4 and R8 should have values roughly equal to half the basic R value (R1, R2, R5, and R6). If standard resistors are not exactly equal to half, you can compensate for it by rounding one of the resistances up to the next standard value, and the other one down to the next standard value.

BANDPASS FILTER

A bandpass filter is somewhat more complicated than a low-pass or a high-pass filter. With those filters you only need to address a single cutoff frequency. If you know the cutoff frequency and the order (slope rate), you pretty much know everything there is to know about the operation of the filter.

A bandpass filter passes all frequency components that fall within a specified band or range. Any frequency component outside this band (either too high or too low) is blocked by the filter circuit. The frequency response of a typical bandpass filter is illustrated in Fig. 5-9. Notice that there are two separate cutoff frequencies. The upper cutoff frequency is at the high end of the passed band and the lower cutoff frequency is at the low end of the passed band.

The action of a bandpass filter can be simulated by a high-pass filter in series with a low-pass filter, as shown in Fig. 5-10. The high-pass filter must have a higher cutoff frequency than the

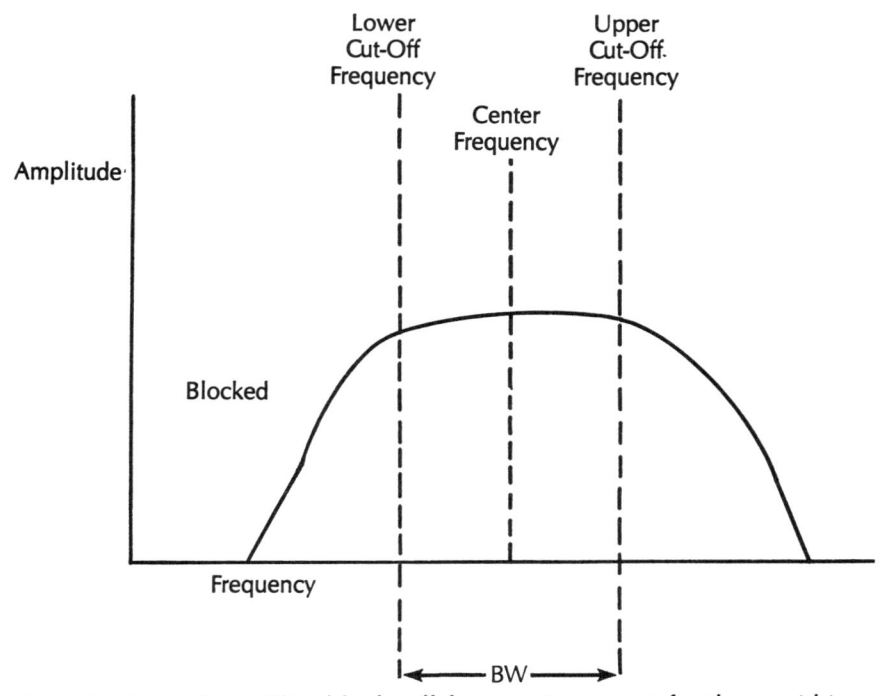

Fig. 5-9 *A bandpass filter blocks all frequencies except for those within a specified band.*

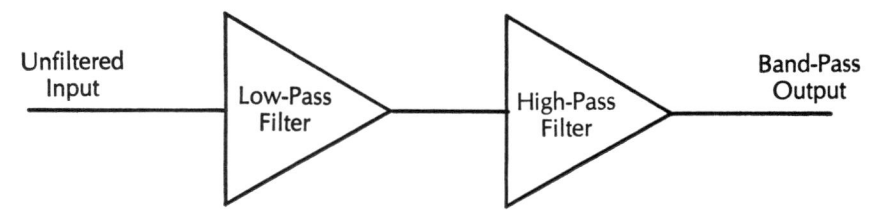

Fig. 5-10 *A bandpass filter can be created from a low-pass filter circuit and a high-pass filter circuit in series.*

low-pass filter. Only those frequency components passed by both filter sections reach the circuit's output. This is illustrated in the composite frequency response graph of Fig. 5-11.

A bandpass filter can be described by identifying its upper cutoff frequency and its lower cutoff frequency, but this is not the usual approach. Generally, bandpass filter circuits are defined by their center frequency. This is the frequency at the exact midpoint of the passed band. The center frequency is halfway between the lower cutoff frequency and the upper cutoff frequency. For example, if the lower cutoff frequency of a certain bandpass filter is 100 Hz, and the upper cutoff frequency is 200 Hz, then the center frequency is 150 Hz.

The center frequency can be derived from the upper and lower cutoff frequencies:

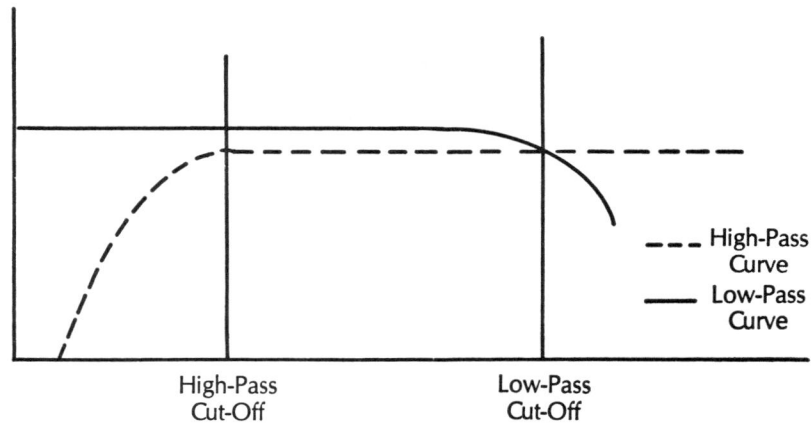

$$F_c = F_l + \frac{F_h - F_l}{2}$$

Fig. 5-11 *This frequency graph illustrates how the composite filter circuit of Fig. 5-10 functions as a bandpass filter.*

F_c is the center frequency, F_l is the lower cutoff frequency, and F_h is the upper cutoff frequency. You can confirm that this formula works by substituting the values from the simple example just given:

$$F_l = 100 \text{ Hz}$$
$$F_h = 200 \text{ Hz}$$
$$F_c = 100 + \frac{200 - 100}{2}$$
$$= 100 + \frac{100}{2}$$
$$= 100 + 50$$
$$= 150 \text{ Hz}$$

Another important specification for a bandpass filter is the bandwidth (*BW*). This is a simple measurement of the size of the passed band or the distance between the upper cutoff frequency and the lower cutoff frequency:

$$BW = F_h - F_l$$

In the example the bandwidth works out to:

$$BW = 200 - 100$$
$$= 100 \text{ Hz}$$

The action of a bandpass filter circuit can be adequately described by either the upper cutoff frequency and the lower cutoff frequency, or by the center frequency and the bandwidth.

Often, a Q specification will be given for a bandpass filter. The Q, or *quality factor* of a filter, is inversely proportional to the bandwidth and to center frequency ratio. A wide band filter circuit has a small Q value and a narrow band filter circuit will have a large Q value. The Q value for any practical filter circuit can be found by dividing the filter's center frequency by the filter's bandwidth:

$$Q = F_c/BW$$

For the example bandpass filter the Q value is:

$$Q = 150/100$$
$$= 1.5$$

This is a fairly typical Q value.

The process can also be worked in reverse. If you know the Q

value and the center frequency of a certain filter circuit, you easily can determine the bandwidth:

$$BW = F_c/Q$$

Once again, you can check this out by plugging in the values from the example:

$$BW = 150/1.5$$
$$= 100 \text{ Hz}$$

A practical bandpass filter circuit is shown in Fig. 5-12. IC1A and its associated components make up a high-pass filter, followed by a low-pass filter comprised of IC1B and its associated components. A suitable parts list for this project is given in Table 5-3. Experiment with alternate component values.

Using the component values given, the lower cutoff frequency is about 300 Hz, while the upper cutoff frequency is approximately 3000 Hz (3 kHz). The bandwidth of this filter is the difference between the two cutoff frequencies:

$$BW = F_h - F_l$$
$$= 3000 - 300$$
$$= 2700 \text{ Hz}$$

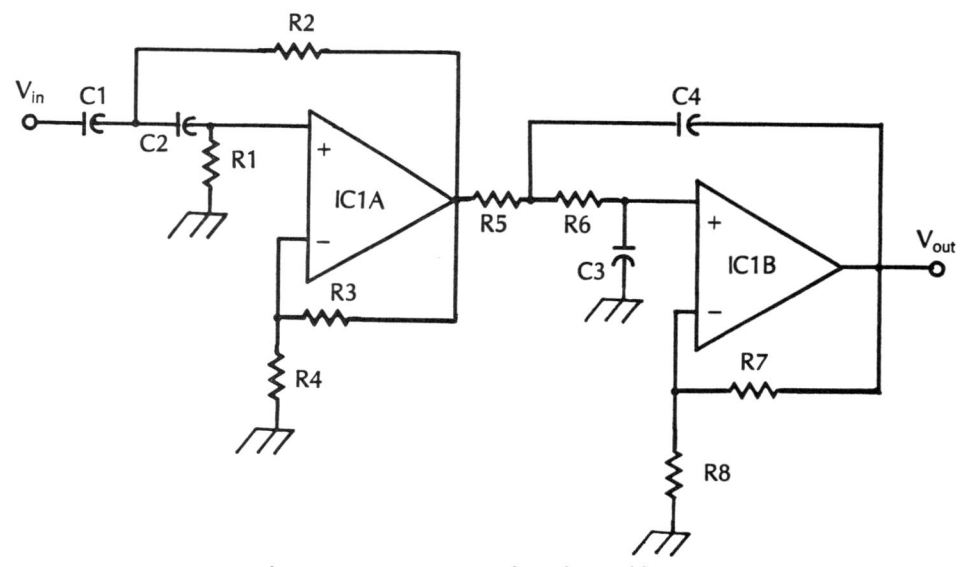

Fig. 5-12 Project 22: bandpass filter.

Table 5-3. Parts List for
Project 22: Bandpass Filter.

Part	Description
IC1	747 dual op amp
C1, C2, C3, C4	0.0068 µF capacitor
R1, R2	100K 1/4-watt 5% resistor
R3, R7	27K 1/4-watt 5% resistor
R4, R8	47K 1/4-watt 5% resistor
R5, R6	10K 1/4-watt 5% resistor

(nominal lower cutoff frequency = 300 Hz)
(nominal upper cutoff frequency = 3000 Hz)

The filter's center frequency in this example works out to:

$$F_c = F_l + \frac{F_h - F_l}{2}$$
$$= 300 + \frac{3000 - 300}{2}$$
$$= 300 + 2700/2$$
$$= 300 + 1350$$
$$= 1650 \text{ Hz}$$

Finally, you can find the filter's Q from the center frequency and the bandwidth:

$$Q = F_c/BW$$
$$= 1650/2700$$
$$= 0.6$$

The component values in the high-pass section are selected to set the lower cutoff frequency, or 300 Hz, in this case. Similarly, the upper cutoff frequency (3000 Hz) is set up by the component values used in the low-pass section of the circuit. The standard filter equation is used for both halves of the circuit:

$$F = \frac{1}{2\pi RC}$$

or substituting for Pi (π):

$$F = \frac{1}{6.28RC}$$

For the high-pass filter section, *R* is the value of resistors R1 and R2, and *C* is the value of capacitors C1 and C2. In the low-pass filter section, *R* is the value of resistors R5 and R6, and *C* is the value of capacitors C3 and C4.

60 HZ NOTCH FILTER

The opposite of a bandpass filter is a band-reject filter. Where a bandpass filter passes only those frequency components within the specified band, a band-reject filter passes all frequency components except those within the specified band.

The frequency response graph of a typical band-reject filter is shown in Fig. 5-13. Notice that the filter circuit in this case essentially takes a bite, or *notch*, out of the frequency spectrum. Because of the appearance of the frequency response graph, the band-reject filter often is referred to as a *notch filter.*

Band-reject filters generally are used to remove specific unwanted frequency components in a system. The bandwidth of

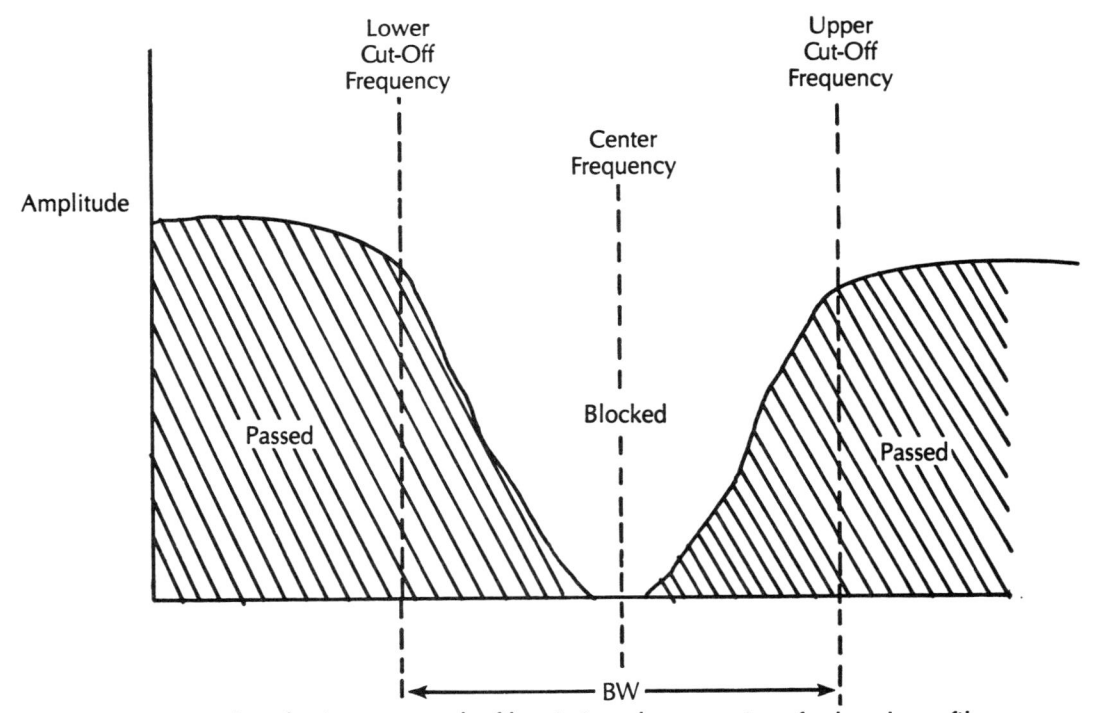

Fig. 5-13 A band reject, or notch, filter is just the opposite of a bandpass filter.

the filter usually will be made very narrow (high Q) to maximize selectivity. In some applications you may need to eliminate a specific frequency, but may want to pass other nearby frequencies.

The most common application for this type of filter is to suppress noise caused by ac power lines. In the United States, household current has a frequency of 60 Hz. Because of the high power levels involved, this 60 Hz signal is radiated out into the atmosphere, and can be picked up by wires or certain components within an electronic circuit. If this strong signal gets into audio equipment an irritating low-pitched hum will be heard. In signal transmission and recording systems, 60 Hz noise could distort, or drown out the desired signal. In such applications a band-reject filter can be used to remove (or significantly reduce) 60 Hz noise.

The action of a band-reject filter can be simulated by a low-pass filter and a high-pass filter in parallel, as shown in Fig. 5-14. Only those frequency components blocked by both filters will be deleted from the circuit's output. If a given frequency component is passed by either the low-pass filter stage or the high-pass filter stage, it will be passed by the band-reject filter circuit as a whole.

A 60 Hz notch, or band-reject filter circuit is illustrated in Fig. 5-15. A suitable parts list for this project is given in Table

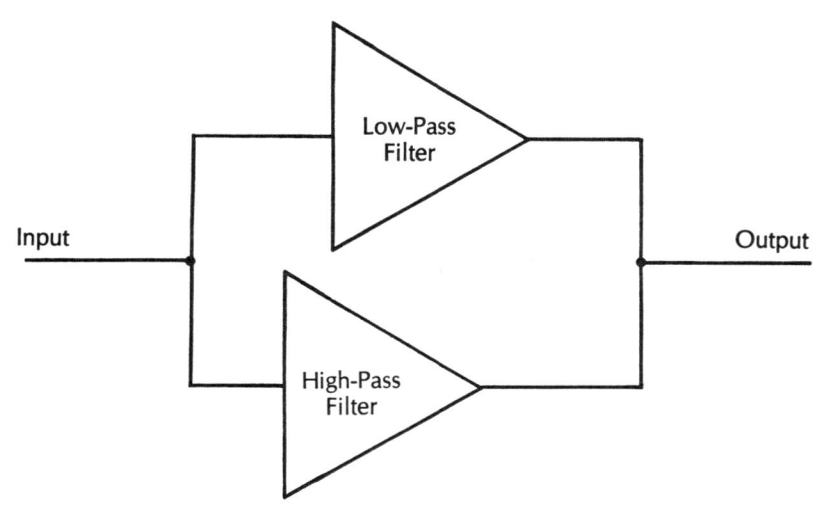

Fig. 5-14 *A notch filter can be created by placing a low-pass filter circuit in parallel with a high-pass filter circuit.*

92 Filter projects

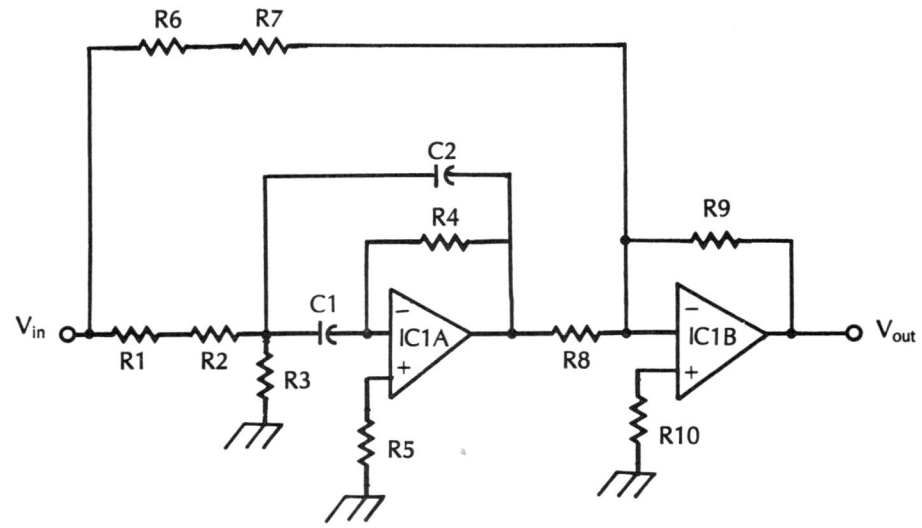

Fig. 5-15 *Project 23: 60 Hz notch filter.*

Table 5-4. Parts List for
Project 23: 60 Hz Notch Filter.

Part	Description
IC1	747 dual op amp
C1, C2	0.22 µF capacitor
R1	2.2K 1/4-watt 5% resistor
R2	22K 1/4-watt 5% resistor
R3	1K 1/4-watt 5% resistor
R4	150K 1/4-watt 5% resistor
R5	120K 1/4-watt 5% resistor
R6	27K 1/4-watt 5% resistor
R7	3.3K 1/4-watt 5% resistor
R8, R9, R10	10K 1/4-watt 5% resistor

5-4. Notice that some components are in series to provide nonstandard values.

STATE VARIABLE FILTER

In some complex systems a *state variable filter* is used. In some ways this type of circuit is analogous to a function generator (see chapter 4). A function generator, you should recall, is an oscillator or signal generator circuit that can produce two or more different waveforms simultaneously. A state variable filter has multiple outputs, each performing a different filtering function. The same key frequency is dominant in each filter section.

Figure 5-16 shows the circuit diagram for a state variable filter. A suitable parts list for this project is given in Table 5-5. You probably will want to experiment with alternate component values in this circuit.

This circuit features a single signal input and three outputs. Each output is a different filtered version of the original input signal, one is a high-pass filter, one is a bandpass filter, and one is a low-pass filter. The same cutoff frequency is used for both the high-pass filter and low-pass filter sections of this circuit. This same frequency is also the center frequency for the bandpass filter section.

The key frequency in this circuit is determined by just four components. Those components are two resistors (R4 and R6), and two capacitors (C1 and C2). Both of these resistors have the

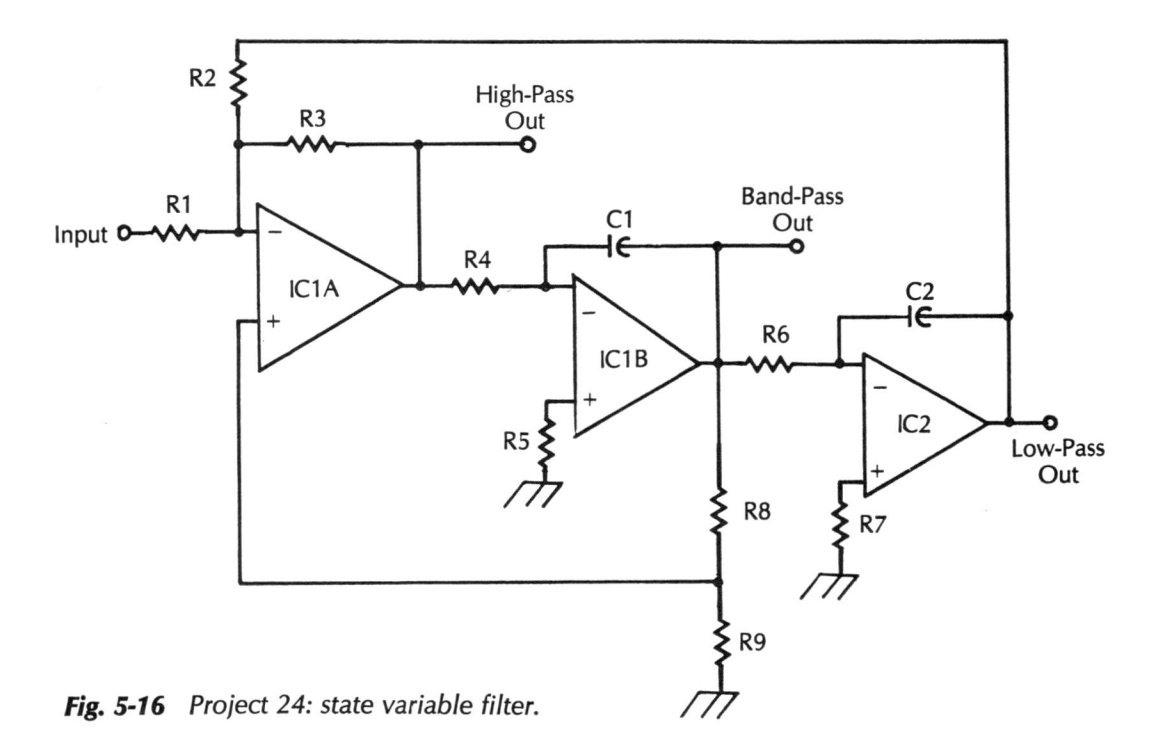

Fig. 5-16 *Project 24: state variable filter.*

Table 5-5. Parts List for Project 24: State Variable Filter.

Part	Description
IC1, IC2	747 dual op amp
C1, C2	0.1 µF capacitor*
R1, R2, R3	10K 1/4-watt 5% resistor
R4, R6	3.3K 1/4-watt 5% resistor*
R5, R7	1K 1/4-watt 5% resistor
R8	120K 1/4-watt 5% resistor**
R9	1K 1/4-watt 5% resistor**

*frequency determining component, see text
**Q determining component, see text

---

same value, as do the two capacitors. That is:

$$R_4 = R_6$$
$$C_1 = C_2$$

These equalities greatly simplify the design equations for this circuit. The circuit probably will not function correctly if these equalities are not met. The key (cutoff/center) frequency for this state variable filter circuit can be found with the following formula:

$$F = \frac{1}{2\pi R_4 C_1}$$

or substituting for Pi (π):

$$F = \frac{1}{6.28 R_4 C_1}$$

The capacitance (C_1) is in farads, and the resistance (R_4) is in ohms. Using the component values suggested in the parts list, the key frequency works out to a value of approximately:

$$F = \frac{1}{6.28 \times 3300 \times 0.0000001}$$
$$= \frac{1}{0.0020724}$$
$$= 482 \text{ Hz}$$

In designing this filter circuit, the aim was for a key frequency of 500 Hz, which was then rounded off to use the nearest standard available component values. The 18 Hz error should be negligible in most applications. In designing a state variable filter, you will want to select the component values to result in a specific key frequency. This is simple enough to do. Just rearrange the basic frequency equation, select a likely capacitor value (C_1), then solve for the required resistance:

$$R_4 = \frac{1}{6.28FC_1}$$

Resistors R8 and R9 set the Q of the bandpass filter section. The Q formula is:

$$Q = 3 + \frac{3R_9}{R_8}$$

In a practical circuit design, select a reasonable (fairly low) value for resistor R9. (1K (1000 ohms) is usually a pretty good choice). Then rearrange the Q equation to solve for resistor R8:

$$R_8 = \frac{3Q-1}{R_9}$$

In most cases you probably will have to round off the value of resistor R8. This will rarely make much noticeable difference in the operation of the circuit.

In selecting the component value in the suggested parts list, a Q of 40 was selected. Using a 1K resistor for R9, the calculated value for resistor R8 worked out to:

$$\begin{aligned}R_8 &= [(3 \times 40) - 1] \times 1000 \\ &= (120 - 1) \times 1000 \\ &= 119 \times 1000 \\ &= 119{,}000\end{aligned}$$

This is close enough to use a standard 120K (120,000 ohms) resistor.

There really isn't too much point in experimenting with the values of resistors R1, R2, R3, R5, and R7. Minor changes in these resistances will not have a noticeable effect on the circuit's operation. If you change any of the resistor values too much, the circuit may not function properly. It is best to keep these resistors at the 10K and 1K values recommended in the parts list.

The signal gain from the high-pass and low-pass outputs will be unity. At the bandpass filter output, the gain at the center frequency will be equal to the Q value (40, using the suggested component values from the parts list). As the signal frequency moves away from the center frequency (in either direction) the gain is reduced. Frequency components outside the filter's band are subjected to negative gain, or attenuation.

All three filter outputs from this circuit are simultaneously available, but if they are used simultaneously, they may tend to interact, and overall circuit performance may be adversely affected. This may or may not be a significant problem, depending on the specific application.

The best high-pass and low-pass performance from this circuit will be achieved with very low Q values (1 or 2, or even less). Most practical bandpass applications will call for a higher Q, so you may be faced with a trade-off in using this circuit.

Three op amp stages are used in this project. If 747 dual op amps are used, an extra op amp section will be left over. It could easily be used in other circuitry, as part of a larger system. Alternately, IC2 could be a single op amp such as the 741.

SPEECH FILTER

In many communications, you primarily are concerned with the speaking voice. In such a system, any low- or high-frequency signal content (within the audible band) will not contain any useful information. Such frequencies effectively will be just noise. Using a bandpass filter to remove the extremes of the audible band will reduce the fidelity, but will improve the understandability of speech. This is the most important factor in a typical communications system.

Most of the information carrying portion of human speech will fall into the 300 Hz to 3000 Hz (3 kHz) range. All frequency components below 300 Hz, or above 3000 Hz can be ignored with no loss of information. A speech filter circuit is shown in Fig. 5-17. A suitable parts list for this project appears as Table 5-6. IC1A and its associated components is a 300 Hz high-pass filter, while IC1B and its associated components form a 3000 Hz low-pass filter. Both filter sections in this circuit exhibit very steep cutoff slopes.

If high-grade, low-noise op amps are substituted for IC1A

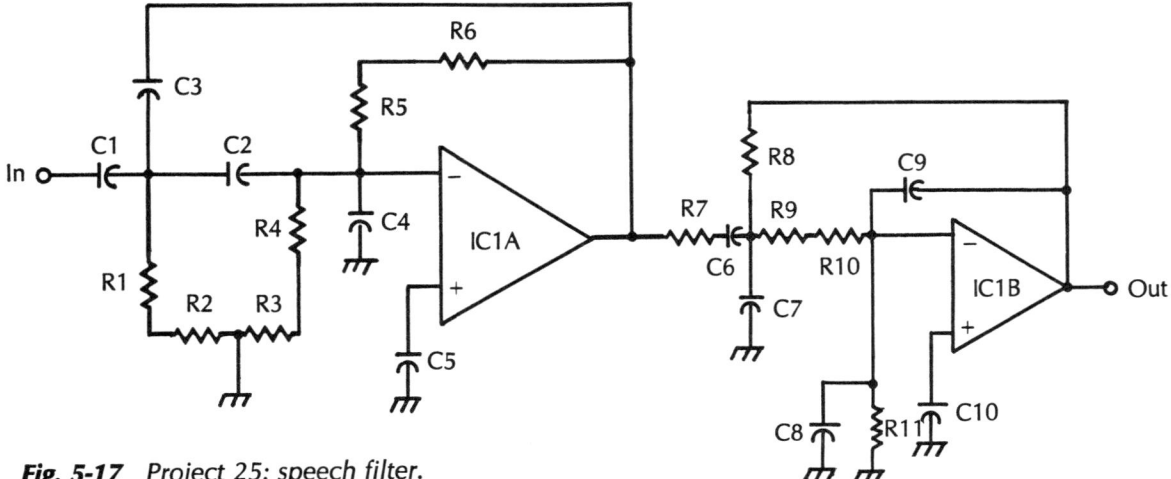

Fig. 5-17 *Project 25: speech filter.*

Table 5-6. Parts List for Project 25: Speech Filter.

Part	Description
IC1	747 dual op amp
C1, C2, C3, C7	560 pF capacitor
C4	0.01 µF capacitor
C5, C6, C10	0.1 µF capacitor
C8	0.0022 µF capacitor
C9	150 pF capacitor
R1	39K 1/4-watt 5% resistor
R2	390K 1/4-watt 5% resistor
R3	220K 1/4-watt 5% resistor
R4	22K 1/4-watt 5% resistor
R5, R6	1M 1/4-watt 5% resistor
R7, R8	270K 1/4-watt 5% resistor
R9	120K 1/4-watt 5% resistor
R10	10K 1/4-watt 5% resistor
R11	47K 1/4-watt 5% resistor

and IC1B, the output signal from this speech filter circuit will be very clear, with a minimum of noise. Noise in the passed band (300 Hz to 3000 Hz) that is added to the signal in the process of transmission or recording/playback will not be removed by this filter. However, the filter circuitry itself will not contribute significantly to the noise level of the output signal.

❖ 6
Test equipment

ANYBODY WORKING IN ELECTRONICS NEEDS TEST EQUIPMENT. THE parameters you are working with in circuits are not directly observable. You can not see a voltage, feel a resistance, or hear a current. You need special devices to convert these qualities into something you can detect with your five senses, such as a light going on, or a pointer moving across a meter. This chapter will present several test equipment circuits using two or more op amps.

DIFFERENTIAL VOLTMETER

Under most circumstances, an unknown voltage will be measured with reference to ground. This is the way most standard voltmeters are designed to work, but this isn't always the best approach. In some cases it isn't even practical.

Some circuits feature floating or balanced voltages which are not referenced directly to ground. In such cases it often will be more convenient and more useful to make the measurements with a differential amplifier. They determine the difference between two voltages which are referenced individually to ground (true zero). A dc differential voltmeter circuit is shown in Fig. 6-1. A suitable parts list for this project is given in Table 6-1.

Five op amp stages are used in this project. IC3 may be a single op amp (such as the 741), or you can use half of a 747 dual op amp chip. The other half of this IC can be used for some other purpose, or it may be left floating.

The two input voltages are applied to IC1A and IC1B. These op amps are connected as noninverting voltage followers to isolate the circuit being tested from the meter circuitry. For maximum accuracy, you might want to substitute high-impedance op

Fig. 6-1 Project 26: differential voltmeter.

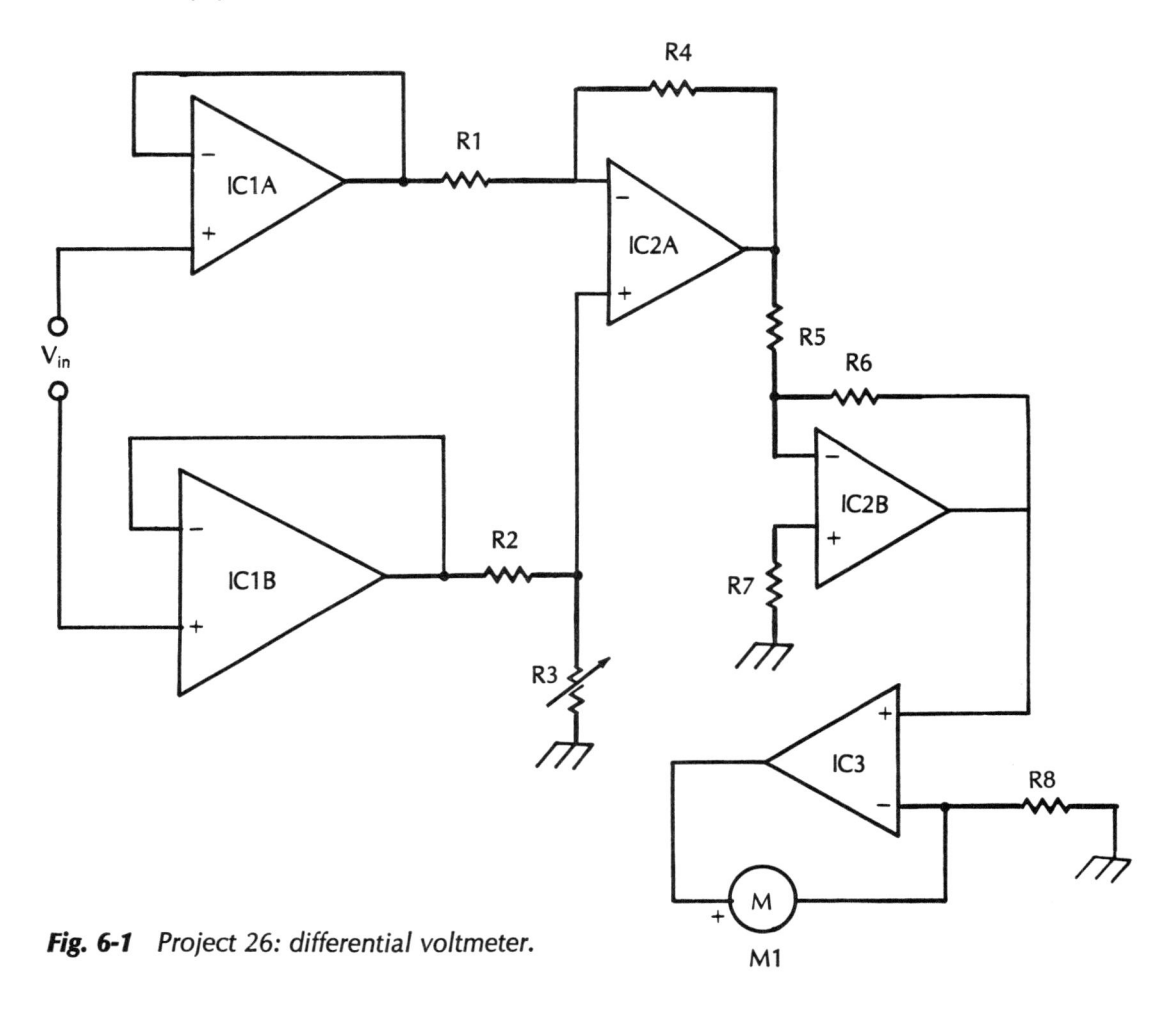

Table 6-1. Parts List for Project 26: Differential Voltmeter.

Part	Description
IC1, IC2	747 dual op amp
IC3	single op amp (741, or 1/2 747)
M1	milliammeter
R1, R2, R5, R6	10K 1/2-watt 5% resistor
R3	100K trimpot
R4	100K 1/2-watt 5% resistor
R7, R8	4.7K 1/2-watt 5% resistor

amps for the 747 (IC1). Op amps with FET or Darlington input stages would be good choices. The 747 op amps will work fine; it's just a question of degree. The choice will depend on just how critical your intended application is. The higher the input impedance, the less loading will be experienced by the circuit under test.

The buffered signals are then fed into an inverting difference amplifier (IC2A). The output signal from this stage is then passed through an inverting amplifier (IC2B). The two inverting stages in series cancel each other out, so the polarity of the output signal from IC2B is the same as the original input polarity.

The final stage (IC3) is a noninverting voltage follower, with a milliammeter (M1) in its feedback path. The meter's internal resistance will have virtually no effect on either the amount of current flowing through it or the read out, because of the circuit arrangement. The meter current is determined solely by the input current being fed into IC3. This results in far more accurate readings across the meter's full range.

This circuit easily can be set up for different measurement ranges by changing the gain of the inverting amplifier (IC2B). If resistors R1 and R2 equal 10K, and resistor R4 equals 100K (as recommended in the parts list), the gain of the difference amplifier stage (IC2A) will be 10. The gain of the entire circuit then can be found with this simple equation:

$$G = \frac{10R_6}{R_5}$$

For the suggested component values from the parts list the circuit gain works out to:

$$G = \frac{10 \times 10{,}000}{10{,}000}$$
$$= \frac{100{,}000}{10{,}000}$$
$$= 10$$

Offset correction resistor R7 should be selected to match the parallel combination of resistors R5 and R6

$$R_7 = \frac{R_5 \times R_6}{R_5 + R_6}$$

This resistor value does not have to be exact. Just use the closest

Fig. 6-2 Project 27: null voltmeter.

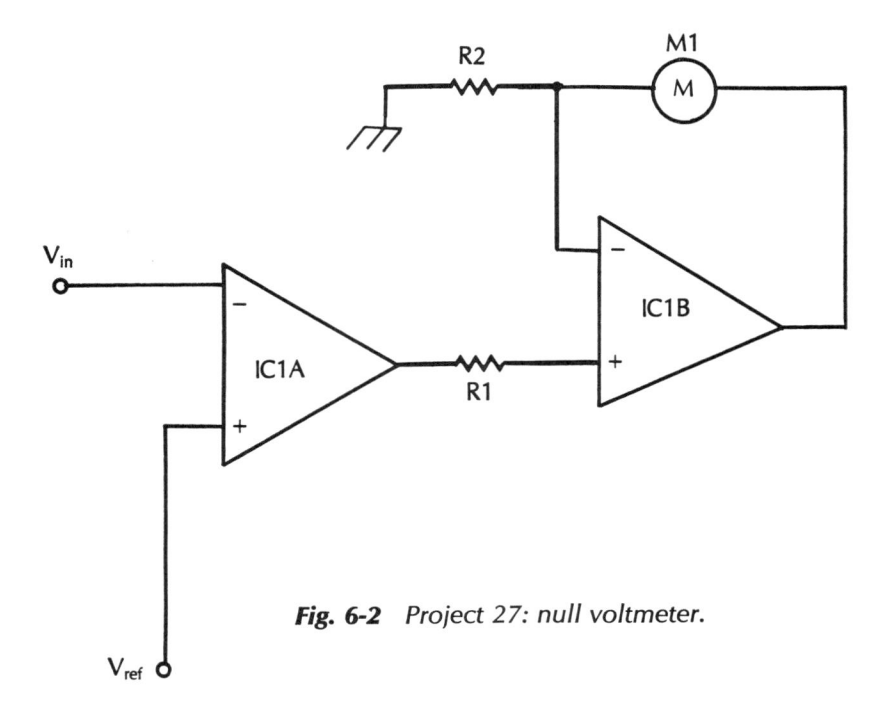

standard resistor value. For example, using 10K resistors for R5 and R6, gives a calculated value of 5K (5,000 ohms) for R7, but a standard 4.7K (4,700 ohms) resistor will work just fine.

NULL VOLTMETER

In some applications it is necessary to tune a voltage to a specific level. This is done most easily with a null voltmeter. This is a device that will tell you if the monitored voltage is above or below the desired level. Zero is the middle of the meter range. This is not necessarily true 0 volts. This point on the meter represents 0 error from a preset reference value. The meter's pointer can move in either direction from this point. A simple op amp null voltmeter circuit is shown in Fig. 6-2. A suitable parts list for this project is given in Table 6-2.

The meter will give a center 0 reading when V_{in} exactly equals V_{ref}. Trimpot R2 is used to calibrate and fine tune the meter. In some applications, this trimpot may be replaced with a simple fixed resistor of an appropriate value.

Table 6-2. Parts List for
Project 27: Null Voltmeter.

Part	Description
IC1	747 dual op amp
M1	−50 mA to +50 mA, center zero milliammeter
R1	10K 1/2-watt 5% resistor
R2	50K trimpot

LED NULL INDICATOR

In some null tuning applications (as discussed in the preceding section) the amount of deviation from center zero isn't particularly important. You just need to know if the measured voltage is too high or too low, and don't care how much it is off. The LED null indicator circuit shown in Fig. 6-3 will come in quite useful for such applications. A suitable parts list for this project appears as Table 6-3.

If the input voltage (V_{in}) exactly equals the reference voltage (V_{ref}), both LEDs (D1 and D2) will be dark (or both may be dimly lit). As V_{in} moves away from V_{ref}, one of the LEDs will start to glow brightly and the other will be completely dark. The farther V_{in} is from V_{ref}, the brighter the appropriate LED will glow

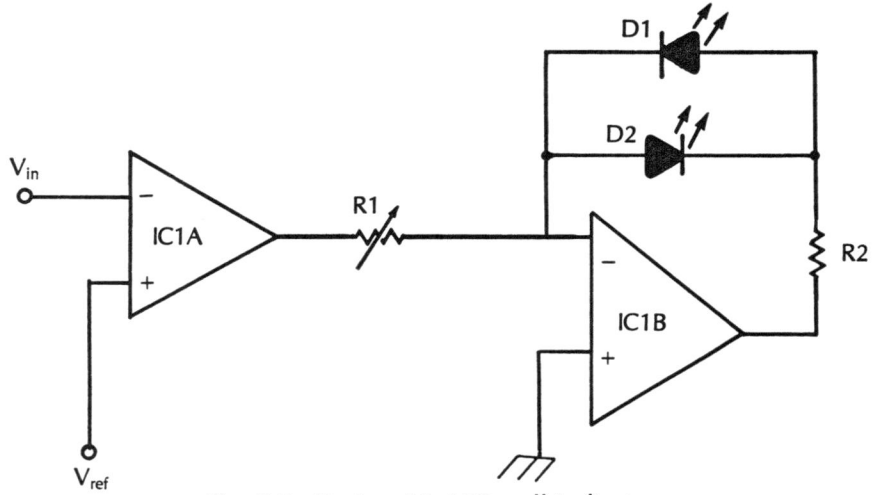

Fig. 6-3 Project 28: LED null indicator.

Table 6-3. Parts List for Project 28: LED Null Indicator.

Part	Description
IC1	747 dual op amp
D1, D2	LED
R1	10K trimpot
R2	220 ohm 1/4-watt 5% resistor

(within the limitations of the device used). If the difference is in the opposite direction (reversed polarity) the other LED will light up and the one that was lit will be extinguished.

Adjust trimpot R1 to fine tune and calibrate the circuit. This compensates for any offset errors within the op amps themselves. The advantages of this LED circuit are that a pair of LEDs is much cheaper and far more durable than a mechanical meter.

DECIBEL METER

Most meter circuits (including those discussed so far in this chapter) make linear measurements. A two volt change in the measured value causes twice the change in the meter reading as a one volt change. In most cases, this is highly desirable and even necessary to get meaningful and consistent results, but this isn't always true. In many applications, like those involving acoustics for example, an exponential scale will give better and more meaningful results. The decibel system was developed for this purpose. A decibel meter circuit is shown in Fig. 6-4. This circuit measures an input voltage along an exponential scale. A suitable parts list for this project is given in Table 6-4.

IC1A and its associated components form a logarithmic amplifier. The output voltage of this stage is equal to the logarithm of the input voltage. (Refer to chapter 2, for more information on logarithmic amplifiers.) Almost any npn transistor can be used for Q1, as long as it can handle the power passing through the circuit. Trimpot R2 is used to calibrate the logarithmic amplifier and compensate for any offsets within the op amp.

IC1B is a noninverting buffer/voltmeter. A milliammeter (M1) is placed in the feedback path of this op amp. Trimpot R3 is used to fine tune the meter's reading for calibration of the circuit.

Fig. 6-4 *Project 29: decibel meter.*

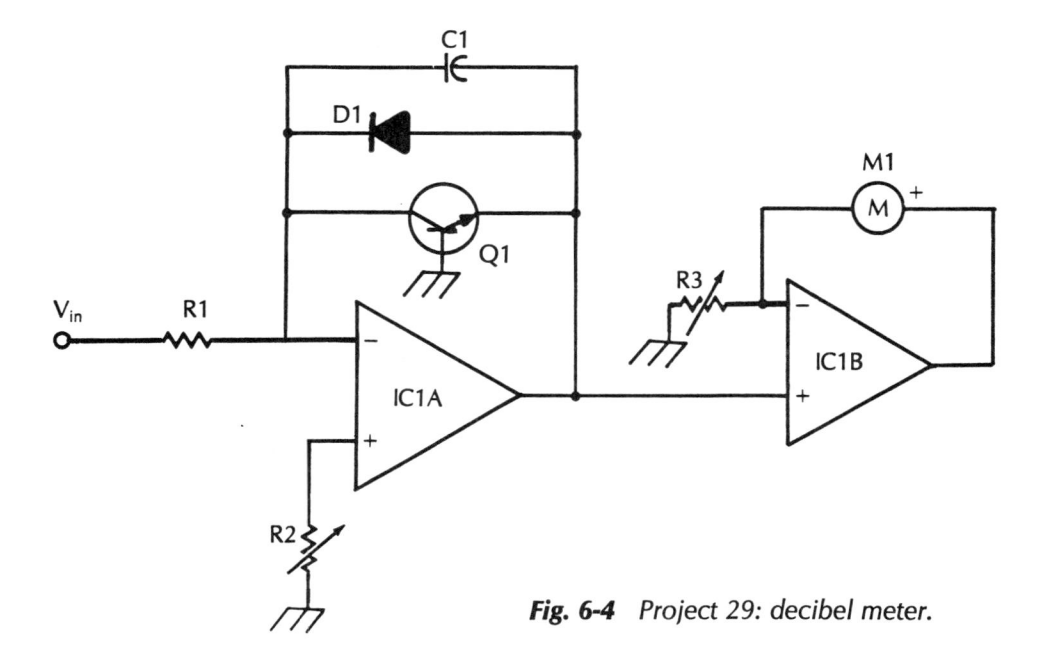

Table 6-4. Parts List for Project 29: Decibel Meter.

Part	Description
IC1	747 dual op amp
Q1	npn transistor (2N2222, or similar)
D1	diode (1N4001, or similar)
M1	100 mA milliammeter
C1	0.1 µF capacitor
R1	47K 1/4-watt 5% resistor
R2, R3	50K trimpot

RMS TO DC CONVERTER

Measuring dc voltages is fairly straightforward. Three volts is three volts. There is no ambiguity. With ac voltages, however, it's a different story. By definition, an ac voltage is continuously changing its value over time.

Just what does 10 volts ac mean? Does it mean that the maximum voltage in the cycle is 10 volts away from the common line (zero volts)? It does if it is 10 volts ac *peak* voltage. It could mean that the cycle passes through a 10 volt range. For example, the positive peak reaches +5 volts, and the negative peak goes as low as −5 volts. The distance from the negative peak to the positive peak is 10 volts. In this it is referred to as 10 volts ac *peak-to-peak*.

The most common and useful way to measure ac voltages is also the least obvious. This method of measuring ac voltages is called the *root-mean-square* (rms) method. The rms voltage can be found with a little math and some averaging. Its usefulness lies in the fact that the rms voltage is the effective equivalent to the same value of dc voltage. That is, a 10 volt ac rms signal produces the same amount of heating capacity as a 10 volt dc signal.

For sine waves the rms voltage is equal to 0.707 times the peak voltage. (This is not true for other ac wave shapes.) For instance, a 10 volt ac peak signal is the same as a 7.07 volt ac rms signal. The power companies supplying house current generate sinusodial (sine wave) signals. This sine waveform is maintained through most power supply circuits until rectification.

Most ac voltmeters read out in rms units. This is the most useful form of measurement, because dc equations can be used with the measured voltages. Ohm's Law applies to ac signals if rms voltages are used.

To be useful, an ac voltmeter must maintain its accuracy over the widest possible frequency range. Standard ac voltmeters are calibrated assuming that the input signal is in the form of a sine wave. They will not give an accurate rms voltage reading for any other waveshape.

One way to solve this problem is to first convert the ac signals into their equivalent dc voltages, which can then be easily and unambiguously measured. The circuit illustrated in Fig. 6-5 is an rms to dc voltage converter that uses the temperature conversion method. The parts list for this project is given in Table 6-5.

Remember that X volts ac rms has the same heating capacity as X volts dc. This circuit makes use of this fact by comparing a pair of heating elements (R1 and R2). R1 and R2 include specialized components known as *thermocouples*. A thermocouple produces an output voltage that is linearly proportional to the temperature (within a specific range).

In this circuit it doesn't matter too much what thermocouples you use, as long as they match. Use the same type for both

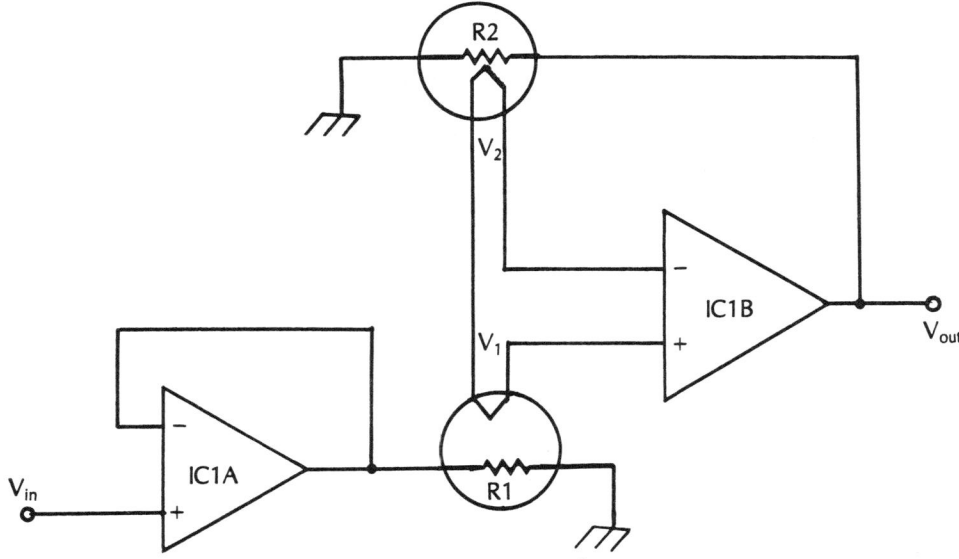

Fig. 6-5 *Project 30: rms-to-dc converter.*

Part	Description
IC1	747 dual op amp
R1, R2	thermocouple

Table 6-5. Parts List for Project 30: Rms-to-dc Converter.

R1 and R2. For the hobbyist, the best bet for finding thermocouples is from a surplus dealer. Because the surplus market is constantly changing, it would be pointless to call for a specific type. It doesn't really matter, as long as the two units are the same.

IC1A is simply a noninverting voltage follower used as a buffer to isolate this conversion circuit from the circuit being measured. The output of this op amp is fed to thermocouple R1. The resistance element is heated by the rms voltage. A dc voltage (V_1) that is proportional to the temperature of R1 is generated by the thermocouple.

This dc voltage is fed into IC1B, which is connected as a difference amplifier with unity gain. The output voltage of IC1B is fed back through the resistance element of thermocouple R2. This thermocouple produces a second dc voltage (V_2) that is proportional to the temperature of its resistance element.

The two thermocouple voltages (V_1 and V_2) are fed to the inputs of the difference amplifier, which will automatically adjust its output until both input voltages are exactly equal. When dc voltage V_2 precisely equals dc voltage V_1, the resistance elements R1 and R2 are at the same temperature. This means that there is the same amount of power across each of the resistance elements. The standard formula for power is:

$$P = \frac{V \times V}{R}$$

or,

$$P = \frac{V^2}{R}$$

Since the two heating elements are at the same temperature, they are using the same amount of power:

$$P_1 = P_2$$

Substituting the appropriate voltage and resistance values gives the following results:

$$V_{in}^2 / R_1 = V_{out}^2 / R_2$$

V_{in} is the input voltage and V_{out} is the output voltage.
You can now algebraically rearrange this equation to solve for the output voltage (V_{out}):

$$V_{out} = \sqrt{(R_2 \times V_{in}^2)/R_1}$$

Assuming the resistance values are equal ($R_1 = R_2$), you can cancel them out, simplifying the formula to:

$$V_{out} = \sqrt{V_{in}^2}$$
$$= V_{in}$$

Remember, V_{in} is the original applied ac voltage in rms, and V_{out} is the dc output voltage of IC1B. They are effectively equal. The output of this circuit can then be measured with a regular dc voltmeter to determine the rms value of the ac signal. This method will work equally well for any ac waveshape, since it is the heating power of the signal being measured, and not the ac voltage itself.

❖ 7
Modulation projects

MODULATORS ARE ELECTRONIC CIRCUITS THAT COMBINE TWO signals so that one is overlayed on the other. Modulation is used most extensively in communications transmissions, such as radio broadcasts. Other applications show up in certain types of recording, data storage, and electronic music.

The two signals in any modulation system are the *program signal* and the *carrier signal*. Some sources may use different names for either or both of these signals. For example, the program signal is often called the *modulating signal*.

The carrier signal is a more or less constant, recurring waveform. Usually the carrier signal is a fairly simple ac waveform, such as a sine wave, or a string of pulses (rectangle wave). Some aspect of the carrier signal is controlled by fluctuations in the program signal. Usually, the carrier signal's frequency will be much higher than the highest frequency included in the program signal. The program signal is the actual information to be transmitted or stored. The program signal might be anything; a speaking voice, music, or digital data.

The modulator circuit combines the program signal with the carrier signal so that the carrier waveform varies in some specific way in step with the program signal. The receiver, or playback device, recreates the original unmodulated carrier signal and compares this with the modulated signal to recover the original program signal.

There are several common types of modulation, each with its own special characteristics. The four most frequently used electronic modulation systems are:

- amplitude modulation (AM)
- frequency modulation (FM)

110 Modulation projects

- pulse amplitude modulation (PAM)
- pulse width modulation (PWM)

In this chapter we will look at projects for all of these modulation systems, except for amplitude modulation.

An amplitude modulation system uses the instantaneous amplitude of the program signal to control the amplitude of the modulated carrier signal. The carrier wave is usually (though not always) a sine wave. In the amplitude modulation process, additional frequency components known as *sidebands* are created. The sideband frequencies appear at the sum and difference points of the program and carrier frequencies.

In an FM system, the instantaneous amplitude of the program signal controls how much the carrier frequency deviates from its nominal (unmodulated) frequency. Once again, the carrier wave is usually (though not always) a sine wave. Sidebands are also created in the frequency modulation process. An FM system usually creates more sidebands than an AM system using the same signals. Standard analog radio broadcasts use either amplitude modulation or frequency modulation.

The remaining pulse modulation systems can be used with either analog or digital program signals. The carrier signal in such systems is always a pulse wave or rectangle wave. The names are pretty self-explanatory. In pulse amplitude modulation, the height (or amplitude) of the carrier pulses is controlled by the instantaneous value of the program signal. In pulse width modulation, the program signal controls the duty cycle, or width (the length of the HIGH time per cycle) of the carrier pulses.

Amplitude modulation systems are the simplest, but they are the most plagued by transmission noise. Straight AM and FM are normally used for analog program signals only. Analog or digital program signals can be transmitted with a PAM or a PWM system.

FM SIGNAL GENERATOR

One of the most popular modulation systems is frequency modulation, or FM. The carrier signal is set at a specific nominal center frequency. Changes in the program signal cause the carrier frequency to deviate from its nominal center value. The circuit

FM signal generator 111

Fig. 7-1 Project 31: FM signal generator.

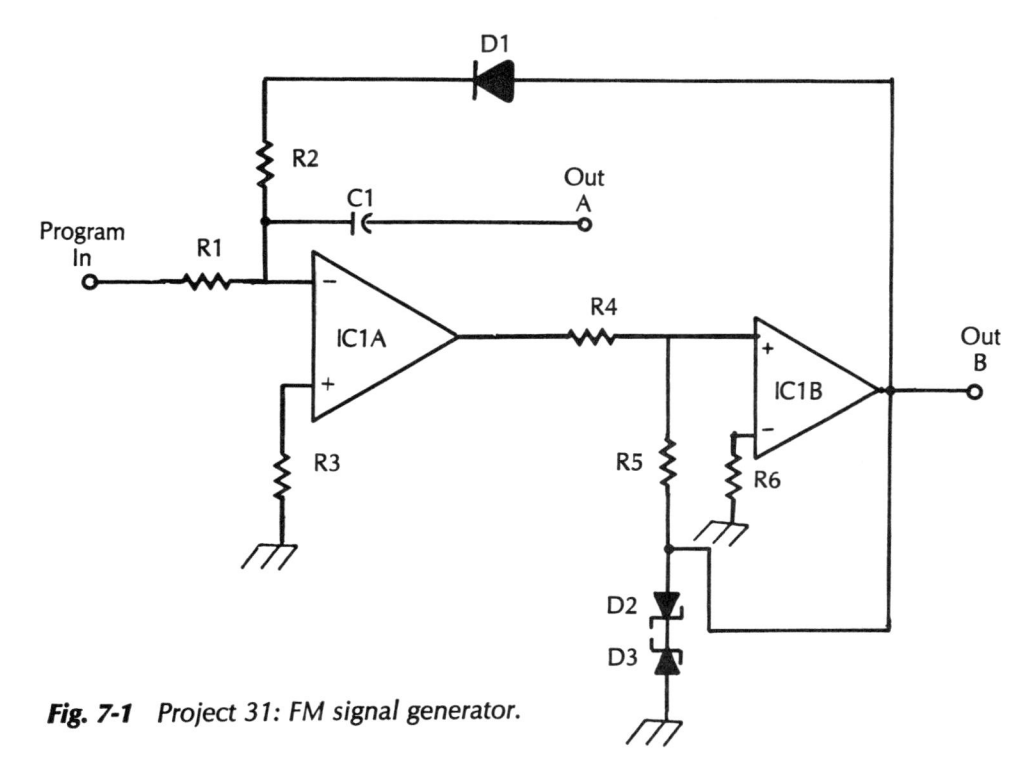

Table 7-1. Parts List for Project 31: FM Signal Generator.

Part	Description
IC1	747 dual op amp
D1	signal diode (1N4148, 1N914, or similar)
D2, D3	6.3-volt zener diode
C1	0.05 µF capacitor
R1, R3	100K 1/4-watt 5% resistor
R2	10K 1/4-watt 5% resistor
R4	82K 1/4-watt 5% resistor
R5, R6	2.2K 1/4-watt 5% resistor

shown in Fig. 7-1 is a frequency modulated function generator. A suitable parts list for this project is given in Table 7-1.

This circuit generates its own carrier signal, and accepts an external input as the program signal. The circuit features two outputs. The signal at output A is an FM sawtooth wave, while

112 Modulation projects

output B puts out an FM rectangle wave. Both output signals operate at the same nominal center frequency, and are equally modulated by the program signal. The only difference between the two outputs is the waveshape of the carrier signal. The waveshape of the carrier signal will also affect the number and strength of the sidebands, because sidebands will be generated for each set of harmonics in the carrier and program signals. As a rule of thumb, the sawtooth wave carrier signal (output A) will exhibit more sidebands than the rectangle wave carrier signal (output B).

For this circuit to function properly, the program signal input should be negative with respect to ground. If possible, the program signal should be kept within the -5 volts to -10 volts range as much as possible.

The output frequency from this circuit is determined by a fairly complex formula:

$$F = \frac{V_{in} R_4}{4 \pi R_1 R_5 C_1 V_z}$$

V_{in} is the program signal voltage at the circuit's input, and V_z is the zener diode voltage, determined by diodes D2 and D3. The resistances (R_1, R_4, and R_5) are in ohms, and the capacitance (C_1) is in farads. Substituting 3.14 for π, the formula can be rewritten as:

$$F = \frac{V_{in} R_4}{12.56 R_1 R_5 C_1 V_z}$$

The parts list for this project suggests the following component values:

R1	100K	(100,000 ohms)
R4	82K	(82,000 ohms)
R5	2.2K	(2,200 ohms)
C1	0.05 μF	(0.00000005 farad)
V_z	6.3 volts	

Using these component values from the parts list, we get a nominal center frequency of:

$$F = \frac{V_{in} \times 82,000}{12.56 \times 100,000 \times 2200 \times 0.00000005 \times 6.3}$$

$$= \frac{V_{in} \times 82{,}000}{870.408}$$
$$= V_{in} \times 114.89 \text{ Hz}$$
$$= V_{in} \times 115 \text{ Hz}$$

If, for example, the instantaneous voltage at the input is −6.5 volts, the modulated frequency at the outputs will be equal to:

$$F = V_{in} \times 115 \text{ Hz}$$
$$= -6.5 \times 115 \text{ Hz}$$
$$= -747.5 \text{ Hz}$$

The negative sign indicates that the signal polarity is inverted by this circuit. That will be of no particular significance in most practical applications, you can ignore the sign in this case. Keeping the program input signal within the −5-volt to −10-volt limits recommended for this circuit, the output frequency will vary from 575 Hz up to 1,150 Hz (1.15 kHz).

Because of the large number of FM generated sidebands, the output signal from this circuit will be very thick and complex. These complex signals can be quite useful in electronic music systems. Use filters to delete any unwanted frequency components from the tone. This FM signal generator project can also be used in signal transmission and storage (recording) systems. Do not use positive voltages at the input of this circuit.

PULSE AMPLITUDE MODULATOR

Pulse amplitude modulation (PAM) is not too dissimilar to regular amplitude modulation (AM). The chief difference lies in the carrier waveform used. In ordinary amplitude modulation, the carrier signal is a sine wave. A sine wave consists of just a fundamental frequency and no harmonics, so the number of sidebands generated is minimized.

In a pulse amplitude modulation system, however, the carrier signal is a string of pulses, or rectangle waves. This makes the signal compatible with digital circuitry, as well as analog circuits. In a pulse amplitude modulator circuit, the instantaneous amplitude of the program signal sets the high level of the current carrier pulse. This process is illustrated in Fig. 7-2.

The pulse amplitude modulator circuit shown in Fig. 7-3 is built around a precision diode. A suitable parts list for this project is given in Table 7-2. Notice that there are two required

114 Modulation projects

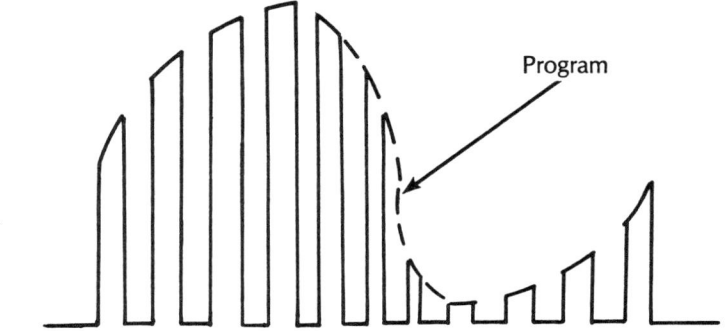

Fig. 7-2 *Pulse amplitude modulation controls the height of the carrier pulses.*

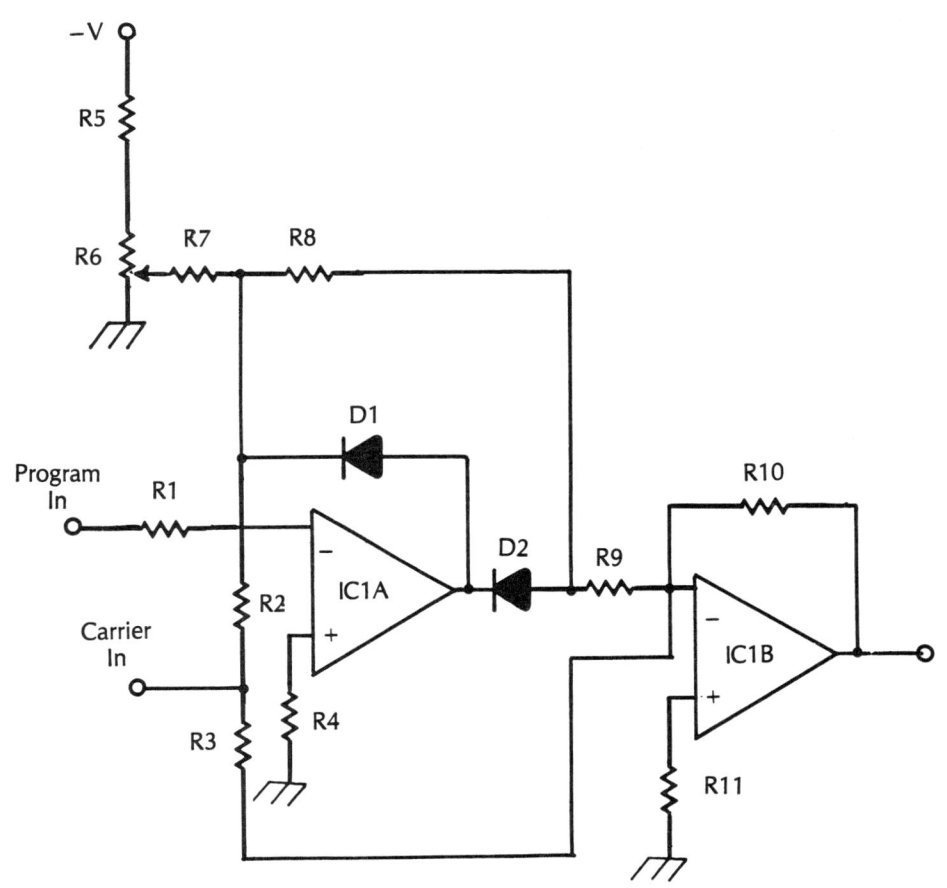

Fig. 7-3 *Project 32: pulse amplitude modulator.*

Table 7-2. Parts List for
Project 32: Pulse Amplitude Modulator.

Part	Description
IC1	747 dual op amp
D1, D2	diode (1N4002, or similar)
R1–R3, R5, R7–R10	10K 1/4-watt 5% resistor
R4	2.7K 1/4-watt 5% resistor
R6	50K trimpot
R11	3.3K 1/4-watt 5% resistor

input signals for this circuit, the program signal and the carrier signal. The program signal may be almost any analog or digital signal, or series of signals. The carrier signal must be a steady frequency rectangle wave.

The program signal may be bipolar in this circuit. That is, it can go positive or negative, above or below zero. Trimpot R6 is adjusted during calibration to set a reference voltage for the circuit. This reference voltage should be made higher (more negative) than the maximum value of the program voltage. That is, the difference between the reference voltage, and the input program voltage must always be negative:

$$V_p - V_{ref} < 0$$

This is assuming the voltage at the carrier input is low. Under these conditions, the precision diode (built around IC1A and its associated components) is reverse biased, blocking the program signal from the output amplifier stage (IC1B and its associated components). The circuit's output remains low until a high carrier plus is fed into the circuit. The high portion of the carrier waveform forward-biases the precision diode network. The program signal plus the carrier pulse now passing through the precision diode is mixed with the original carrier signal in the two input output amplifier stage (IC1B). Compare this portion of the circuit to the audio mixer project presented in chapter 3.

The program signal is inverted by IC1A and reinverted back to its original polarity by IC1B. The carrier pulse is fed directly into the inverting input of the output amplifier stage. The inverted carrier pulse cancels out the noninverted carrier pulse.

The output of amplifier IC1B is the original program input signal, but only during the high portion of the carrier signal. During the low portion of the carrier wave, the program signal is cut off, and the circuit's output is simply 0, or low. In effect, you now have the response illustrated back in Fig. 7-2.

A modulation system isn't much good unless you have some way to recover the original program signal. This process is known as *demodulation*. (An exception is in electronic music applications, where only the modulated signal is of interest.)

Figure 7-4 shows a simplified block diagram for a PAM demodulation circuit. The switch driver stage is activated by each signal pulse, acting as a clock for the sample-and-hold stage. The output of the sample-and-hold is a staircase wave, equal to the sum of the original program signal and a local reference voltage (V_{ref}). The reference voltage may be either positive or negative. The V_{ref} voltage should be the opposite polarity of the sample-and-hold offset. By mixing this polarity-reversed reference voltage with the main signal, the reference voltage offset is canceled out. The output of the amplifier is a duplication of the original program signal.

The capacitor (C1) in the amplifier's feedback loop serves as a simple low-pass filter, to smooth out the recovered signal. In this circuit, the output's polarity is inverted from the original

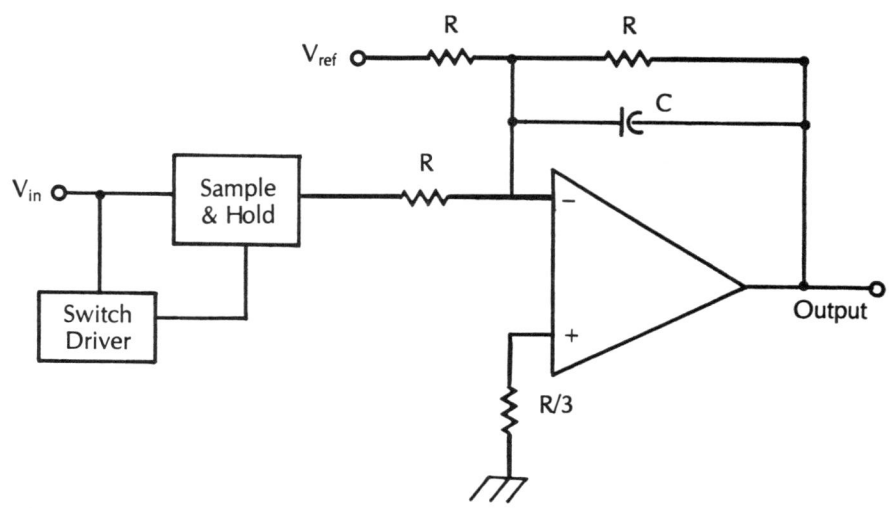

Fig. 7-4 This type of circuit can demodulate pulse amplitude modulated signals.

input signal back at the modulator circuit. In most applications, this won't be significant. In many cases, the effect won't even be detectable. However, in an oddball application requiring correct polarity of the program signal, you can simply pass the output of this signal through an inverting voltage follower stage to reinvert the signal polarity back to its original state.

PULSE WIDTH MODULATOR

An increasingly popular modulation system is pulse width modulation (PWM). A string of equal amplitude rectangle wave pulses act as the carrier. The length of each pulse (or the pulse width) is determined by the instantaneous value of the modulating program signal. Another way to describe this process is that the duty cycle of the modulated output waveform varies in step with the program signal.

Pulse Width Modulation can be used to transmit or store either analog or digital data. A Pulse Width Modulator circuit built around a pair of op amps is shown in Fig. 7-5. A suitable parts list for this project is given in Table 7-3. The zener diodes are selected to determine the amplitude of the pulses at the circuit's output.

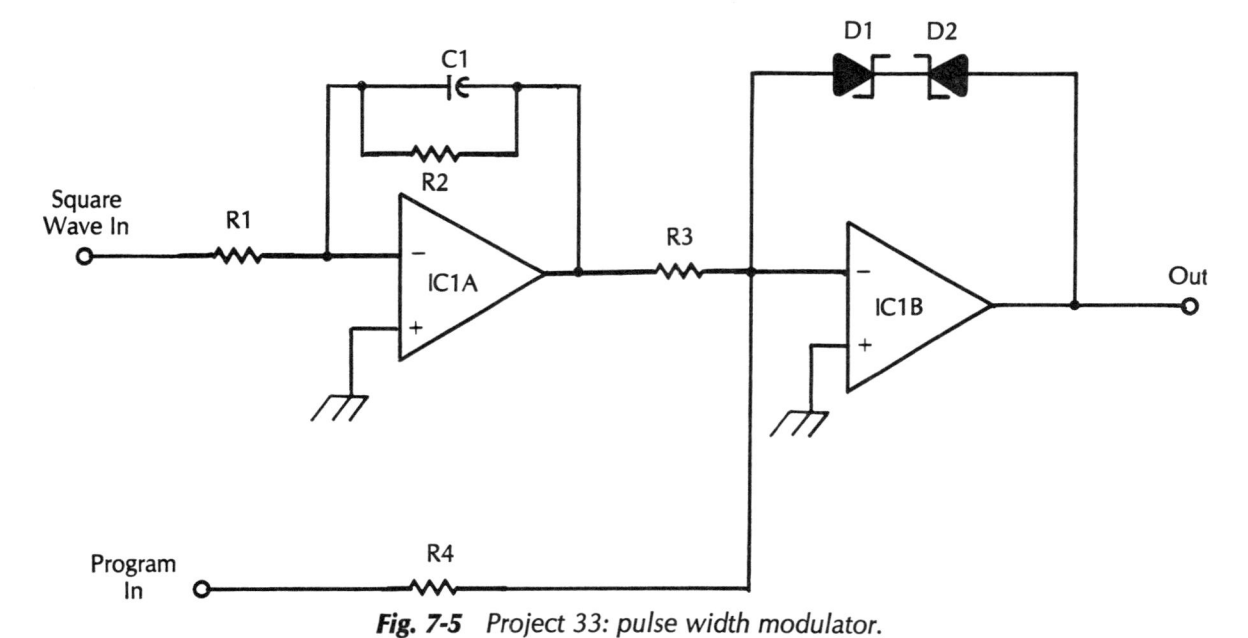

Fig. 7-5 *Project 33: pulse width modulator.*

118 Modulation projects

**Table 7-3.
Parts List for Project 33:
Pulse Width Modulator.**

Part	Description
IC1	747 dual op amp
D1, D2	6.8-volt zener diode
C1	0.01 µF capacitor
R1	27K 5% 1/4-watt resistor
R2	1M 5% 1/4-watt resistor
R3, R4	10K 5% 1/4-watt resistor

Once a PWM signal has been transmitted or stored, the receiver will obviously need some way to demodulate the PWM signal to recover the original program signal. Since the amplitude of the carrier pulses remains constant, this isn't difficult at all. Simple low-pass filtering at the receiver end will produce an output voltage that is proportional to the instantaneous pulse width. Since the pulse width corresponds to the original program signal, it naturally follows that the output of the low-pass filter will resemble the signal originally fed into the program input of the pulse width modulator circuit.

❖ 8
Pulse circuits

PULSE CIRCUITS ARE BECOMING INCREASINGLY IMPORTANT IN THE field of modern electronics. This is largely because of the increased emphasis on digital circuitry. Pulse circuits can often cross the line between digital and analog circuitry.

In a pulse circuit, the signal has only two possible states, low and high. Theoretically, the circuit will switch instantly from one state to the other. In reality there will be a finite transition time between states, but it will be very short and can be reasonably ignored in all but the most critical of applications. In operational amplifier circuits the transition time will be limited by the device's *slew rate*. The slew rate of an op amp is a measurement of how rapidly the output signal can respond to large, very fast changes in the input signal. Refer to chapter 1 for more information on op amps and their slew rate specification.

The basic pulse circuit is known as a *multivibrator*. There are three types of multivibrator circuits: the monostable multivibrator, the bistable multivibrator, and the astable multivibrator. As you shall see, these three variations cover all possible combinations.

A monostable multivibrator has one stable state. The opposite state is unstable. Normally, the circuit's output will remain in its stable state. When an external trigger pulse is detected at the input of the circuit, the output jumps to the opposite, unstable state for a specific, predetermined period of time. After this timing period, the circuit's output reverts back to its normal, stable state. There is one output pulse for each input (*trigger*) pulse. For this reason, this type of circuit is often called a *one-shot* timer.

The output pulse has a fixed duration, dependent on specific resistor and capacitor values within the monostable multivibrator circuit. The length of the input (trigger) pulse has no effect on the length of the output pulse. Because the output pulse lasts for a specific, repeatable time period, monostable multivibrators also are called *timers*.

Another popular name for the monostable multivibrator circuit is the *pulse stretcher*. This is because in most applications, the output pulse will be longer than the trigger pulse. A short pulse is effectively made longer, or is stretched out. The operation of a typical monostable multivibrator circuit is illustrated in Fig. 8-1.

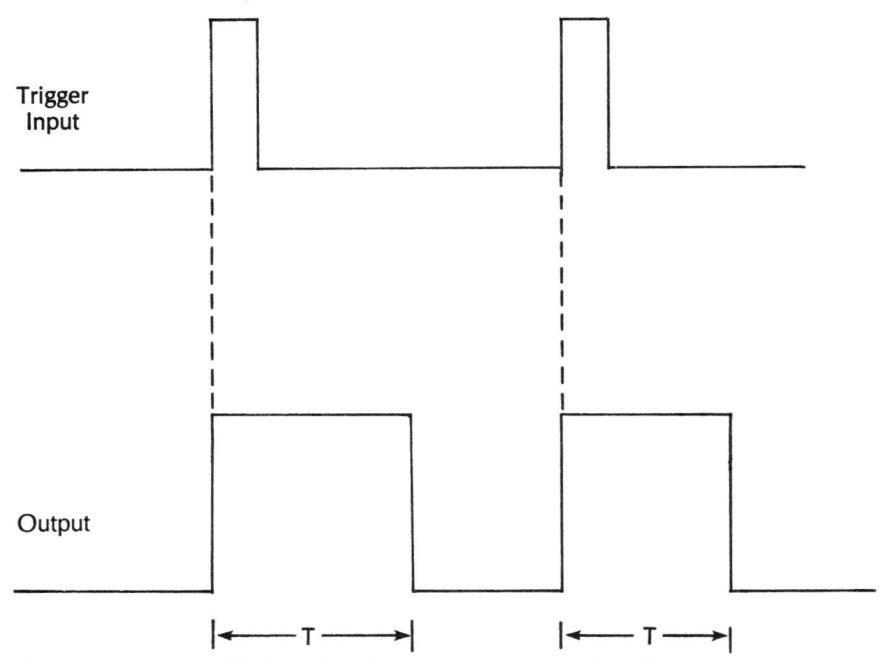

Fig. 8-1 A monostable multivibrator puts out one fixed-length output pulse for every input trigger pulse.

A bistable multivibrator circuit has two stable states. Either output state (high or low) can be held indefinitely (as long as power is continuously applied to the circuit, of course). Like the monostable multivibrator, the bistable multivibrator has a trigger input. Each time a trigger pulse is detected at the circuit's input, the circuit output reverses its state, going from high to low, or from low to high. Because the output flip-flops back and forth between states, the bistable multivibrator is also widely known as the *flip-flop*.

The bistable multivibrator circuit is a very simple one-bit memory. As long as power is continuously applied to the circuit, it remembers the last state it was triggered into. If a continuous stream of pulses are fed into the trigger input of a bistable multivibrator circuit, the output will be a rectangle wave with exactly one-half the frequency of the original input signal. This happens because there is one complete output cycle for every two complete input cycles.

To see how this works, assume that the output starts out in the low state. Also assume that this particular bistable multivibrator circuit is triggered by a low to high transition. (Some bistable multivibrator circuits are triggered by a high to low transition.) Now see what happens on each successive input (trigger) pulse.

Input Pulse	Input State	Output State	Output Pulse
0	Low	Low	0
1	High	High	1
	Low	High	
2	High	Low	
	Low	Low	
3	High	High	2
	Low	High	
4	High	Low	
	Low	Low	
5	High	High	3
	Low	High	
6	High	Low	
	Low	Low	

As you can see there are two complete input cycles for every complete output cycle. The operation of a typical bistable multivibrator circuit is illustrated in Fig. 8-2.

Most bistable multivibrator circuits have two outputs. The main output is labeled Q, and the secondary output is labeled $\overline{Q}$ (Not Q). The Q output is always at the opposite state as the $\overline{Q}$ output. That is, if Q is high, then $\overline{Q}$ is low, and vice versa.

The astable multivibrator circuit has no stable states. The output continuously switches back and forth between the low and high states at a regular rate (determined by specific resistor and capacitor values within the circuit). Unlike the monostable and bistable multivibrators, the astable multivibrator generally

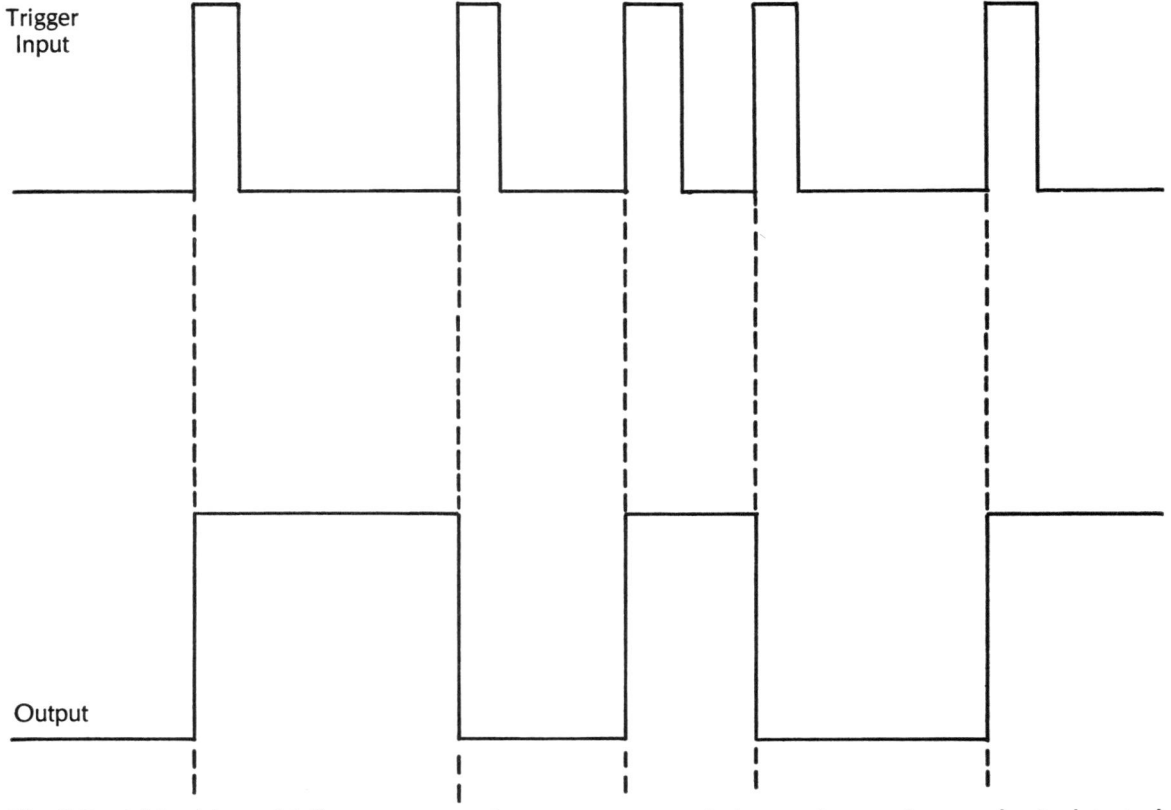

Fig. 8-2 *A bistable multivibrator reverses its output state each time an input trigger pulse is detected.*

does not have a trigger input. Most astable multivibrator circuits are self-starting. As soon as power is applied to the circuit, the output oscillates back and forth between its two available states.

The operation of a typical astable multivibrator circuit is illustrated in Fig. 8-3. As you can plainly see this is basically just a rectangle wave generator. (For more information on signal generator circuits, refer to chapter 4.)

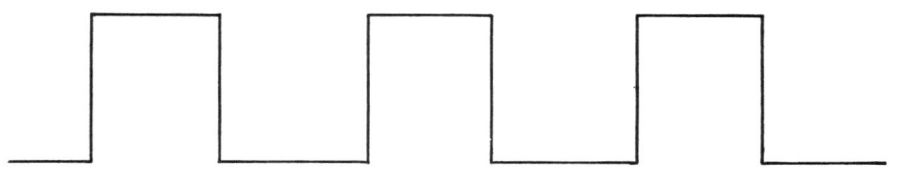

Fig. 8-3 *An astable multivibrator continuously switches back and forth between output states.*

ONE-SHOT TIMER

Operational amplifiers can be used in all three of the basic types of multivibrator circuits. The following projects illustrate their use.

ONE-SHOT TIMER

A practical *monostable multivibrator,* or one-shot timer circuit, built around a pair of op amps is shown in Fig. 8-4. A suitable parts list for this project is given in Table 8-1. You may want to experiment with alternate component values in this circuit, especially for capacitor C1 and resistor R5. These two components set the timing period of the monostable multivibrator.

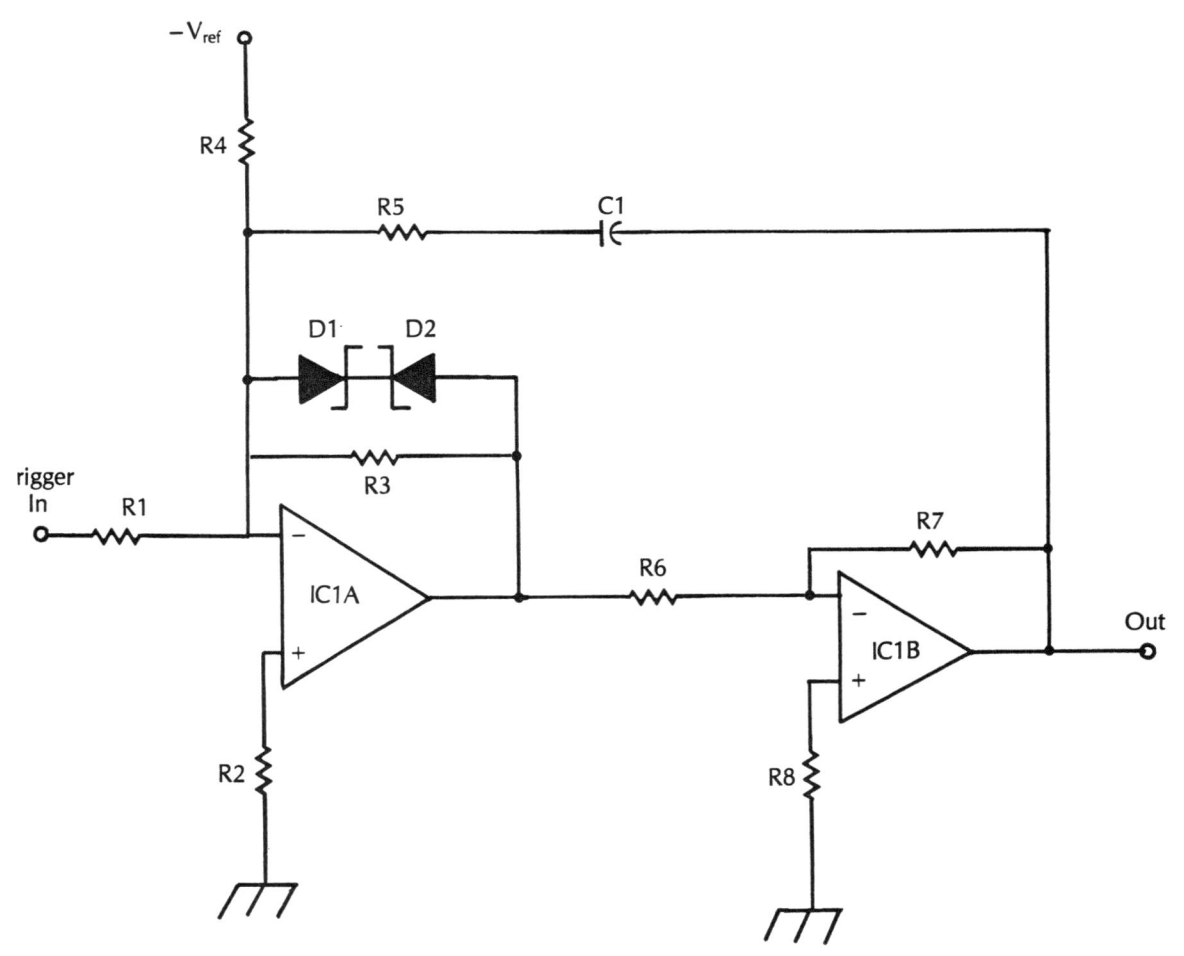

Fig. 8-4 *Project 34: one-shot timer.*

Table 8-1. Parts List for Project 34: One-Shot Timer.

Part	Description
IC1	747 dual op amp
D1, D2	zener diode (5.1 volt)
C1	0.5 µF capacitor (see text)
R1, R6, R7	10K 1/4-watt 5% resistor
R2	68K 1/4-watt 5% resistor
R3	1M 1/4-watt 5% resistor
R4	270K 1/4-watt 5% resistor
R5	470K 1/4-watt 5% resistor (see text)
R8	4.7K 1/4-watt 5% resistor

Op amp IC1A is a summing amplifier, combining the trigger input, a feedback signal (through capacitor C1 and resistor R5), and a negative reference voltage. The two zener diodes (D1 and D2) in the feedback path of this op amp limit the voltage swings at its output.

IC1B is a unity gain inverting amplifier, creating a positive voltage that can be fed back to the circuit's input through the timing components (resistor R5 and capacitor C1). Unity gain in this stage is achieved by giving resistor R6 the same value as resistor R7. For maximum stability, the value of resistor R8 should be half that of R6 or R7.

In operation, when there has been no trigger pulse for awhile (so the circuit is in its normal, stable condition) the negative reference voltage forces the output of op amp IC1A to its maximum positive value, as defined by the zener diodes. You can call this voltage $+V_z$. The second op amp stage (IC1B) inverts this signal with unity gain, so the circuit's output at this point is $-V_z$. The capacitor is fully charged with the end adjacent to resistor R5 positive.

To change the output state of this circuit, it is necessary to apply a trigger pulse of sufficient voltage to overcome the sum of the feedback voltage through capacitor C1 and resistor R5; and the negative reference voltage. The output of IC1A goes negative $(-V_z)$, so the output of IC1B (the circuit output) goes positive $(+V_z)$, or high.

This condition is inherently unstable. The capacitor (C1) starts to discharge through the discharge resistor (R5). The voltage through this resistor decays towards zero. At some point, the negative reference voltage will again take over as the strongest control signal, and force the output of IC1A back to $+V_z$. This brings the circuit's output back to the normal, stable low state. The values of resistor R5 and capacitor C1 have a direct effect on the timing period of this monostable multivibrator circuit, as do several of the other resistor values, and the negative reference voltage.

DC COUPLED BISTABLE MULTIVIBRATOR

Figure 8-5 shows a practical bistable multivibrator circuit using the 747 dual op amp IC. A suitable parts list for this project appears as Table 8-2.

This circuit is dc coupled. The trigger input is fed directly into the circuit through resistor R1.

A feedback voltage is set up through resistors R2 and R3. The ratio of the values of these two resistors determines the feedback voltage seen by the op amp. It might be necessary to experiment

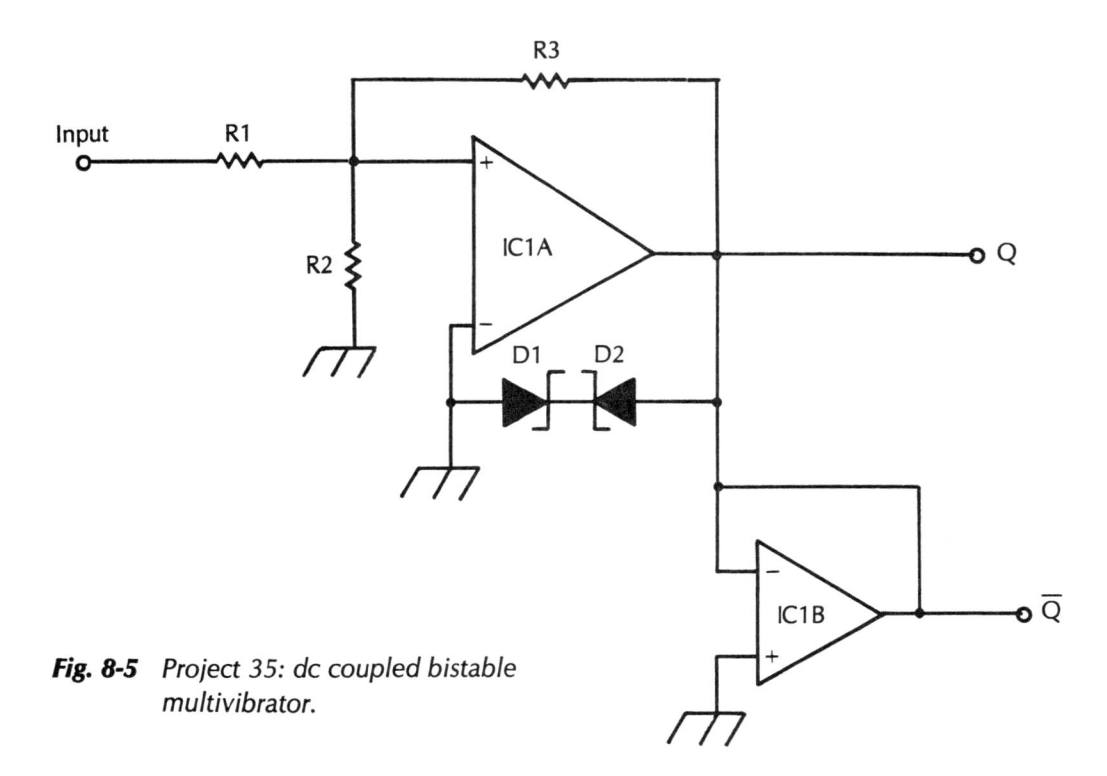

Fig. 8-5 *Project 35: dc coupled bistable multivibrator.*

126 Pulse circuits

**Table 8-2.
Parts List for Project 35:
DC Coupled
Bistable Multivibrator.**

Part	Description
IC1	747 dual op amp
D1, D2	zener diode (5.1 volt)
R1	4.7K 1/4-watt 5% resistor
R2	10K 1/4-watt 5% resistor
R3	22K 1/4-watt 5% resistor

with different resistor values in this circuit to provide the best and most reliable performance with the desired trigger signal.

A positive going input (trigger) pulse forces the Q output of this circuit negative (low), while a negative going input (trigger) pulse drives the circuit's Q output positive (high). The voltage limits of the output signal are set by the zener diodes (D1 and D2). IC1B is a simple inverting voltage follower, to provide the $\overline{Q}$ (Not Q) output. The output of IC1B ($\overline{Q}$) will always be at the opposite state as the output of IC1A (Q).

AC COUPLED BISTABLE MULTIVIBRATOR

This project is similar to the preceding one, except this bistable multivibrator circuit is ac coupled, rather than dc coupled (as in Fig. 8-5). The circuit for this project is shown in Fig. 8-6 and the parts list is given in Table 8-3.

The trigger pulse inputs are ac coupled through capacitor C1. The voltage limits of the output signal are set by the zener diodes (D1 and D2).

A feedback voltage once again is set up through resistors R2 and R3. The ratio of the values of these two resistors determines the feedback voltage seen by the op amp. It may be necessary to experiment with different resistor values in this circuit to provide the best and most reliable performance with the desired trigger signal.

Unlike the preceding project, only positive trigger pulses are used with this circuit. No negative trigger pulses are required. This bistable multivibrator circuit is triggered by the transitions in the input trigger signal. Assuming that the circuit's main out-

Ac coupled bistable multivibrator

Fig. 8-6 *Project 36: ac coupled bistable multivibrator.*

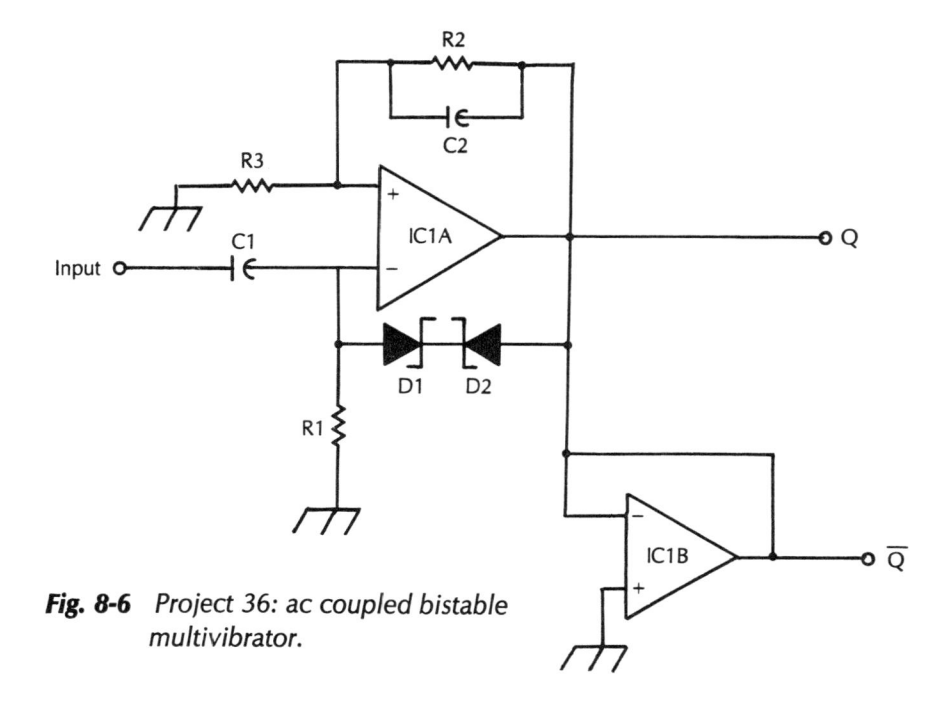

Table 8-3. Parts List for Project 36: AC Coupled Bistable Multivibrator.

Part	Description
IC1	747 dual op amp
D1, D2	zener diode (5.1 volt)
C1	0.01 µF capacitor
C2	0.1 µF capacitor
R1	4.7K ¼-watt 5% resistor
R2	10K ¼-watt 5% resistor
R3	33K ¼-watt 5% resistor

put (Q) is high, the leading edge (low to high transition) of the next pulse at the trigger input will force the main (Q) output to its low state. On the next received trigger pulse, the circuit will be activated by the trailing edge (high to low) transition, driving the main (Q) output into the high state again. Alternate edges of the trigger pulses are used to switch the alternate output states.

Capacitor C2 is included in the circuit to slow down its response a bit. Without this capacitor, when the leading edge of a given trigger pulse drives the Q output low, the trailing edge of the same pulse could conceivably cause the circuit to prematurely switch back into its high Q output state.

In some applications, it may be necessary to experiment with different values for capacitor C2 to prevent such false triggering problems. You should be aware that the value of this capacitor interacts with the values of the resistors in the circuit, and changing the capacitance might alter the necessary feedback voltage (set by the ratio of resistors R2 and R3) to permit reliable triggering.

IC1B is a simple inverting voltage follower, to provide the $\overline{Q}$ (Not Q) output. The output of IC1B ($\overline{Q}$) will always be at the opposite state as the output of IC1A (Q).

ASTABLE MULTIVIBRATOR

An astable multivibrator circuit using two operational amplifiers is illustrated in Fig. 8-7. Compare this circuit with the mono-

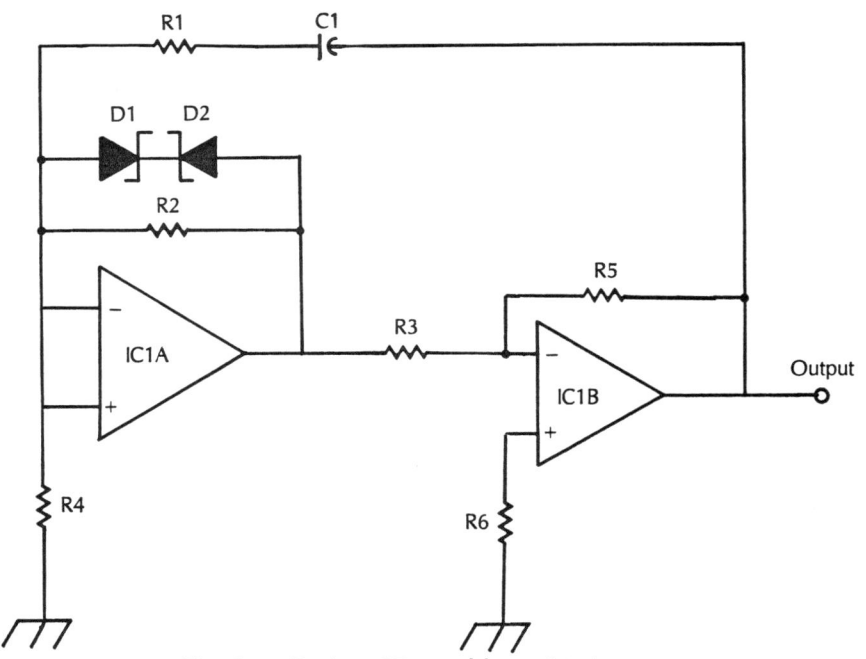

Fig. 8-7 *Project 37: astable multivibrator.*

stable multivibrator circuit shown back in Fig. 8-4. This circuit is a self-triggering monostable multivibrator. Each time the timing period is concluded the circuit's output is automatically forced into the opposite state. This back and forth action continues as long as power is applied to the circuit. A suitable parts list for this project is given in Table 8-4.

The timing of this circuit is controlled primarily by the values of resistor R1 and capacitor C1. Be sure to experiment with alternate values for these two components. None of the other component values in this circuit are terribly critical, so the most worthwhile experimentation will be with resistor R1 and capacitor C1.

The same timing period is used for both the high portion and the low portion of the cycle. The output of this circuit will be a square wave with a duty cycle of 1:2. A true square wave consists of just the fundamental and all odd harmonics.

Because of limitations of the linearity in this circuit, the output is not quite a true square wave. If you observe this signal closely on an oscilloscope, you will see that the tops and bottoms of the waveform are somewhat sloped, rather than perfectly flat, as in a true square wave. Fortunately, this inaccuracy won't be very important in most applications.

There are some important exceptions. One of the more common applications of a square wave generator is to provide a test signal to determine the linearity of an amplifier circuit (or other similar circuitry). If the original square wave isn't extremely linear, the test results will be of very limited value.

Table 8-4. Parts List for Project 37: Astable Multivibrator.

Part	Description
IC1	747 dual op amp
D1, D2	zener diode (5.1 volt)
C1	0.01 μF capacitor (see text)
R1	100K 1/4-watt 5% resistor (see text)
R2	22K 1/4-watt 5% resistor
R3, R4, R5	10K 1/4-watt 5% resistor
R6	4.7K 1/4-watt 5% resistor

COMPARING DIGITAL AND ANALOG CIRCUITRY

Operational amplifiers are analog devices, and are normally used in analog, or *linear*, circuitry. Parameters (such as voltage and current) in an analog circuit can vary over a continuous range. There is always at least one intermediate value between any two values. For example, between 2 volts and 3 volts, there is 2.5 volts. Between 2.5 volts and 3 volts, there is 2.75 volts. Between 2.75 volts and 3 volts, there is 2.875 volts. This splitting process can be continued infinitely, or at least to the limits of measurability.

More and more electronic circuits these days are digital, rather than analog. In a digital circuit, discrete, numerical values are used. Intermediate values are not allowed.

In everyday life, you use the decimal numbering system, with ten digits. In digital electronics, the binary system is used. There are just two digits (1 and 0). There can never be an intermediate value between 0 and 1 in any digital circuit.

The use of the binary numbering system in digital electronics is not an arbitrary choice. As it happens, it is very easy to unambiguously represent the two digits of the binary numbering system by electrical means. Generally a low voltage (typically near zero) is used to represent a binary 0, and a high voltage (typically close to the circuit's supply voltage) represents a binary 1.

Binary values in electronic circuits are often referred to as digital logic. A single digit (a 0 or a 1) in the binary numbering system is known as a *bit* (BInary digIT). By itself, a single bit isn't good for very much. But multiple bits can be combined to represent larger values.

In the decimal numbering system you are used to, when you need to represent a value larger than the highest available digit (9), you add another column to the left. The digit value in this column is multiplied by the number of available digits, or ten. Thus:

$$23 = (2 \times 10) + 3$$

This process can be expanded as far as necessary. Each additional new column to the left is raised by another power of ten (the base of the numbering system). For example:

$$41754 = (4 \times 10^4) + (1 \times 10^3) + (7 \times 10^2) + (5 \times 10^1) + (4 \times 10^0)$$

Remember that any value raised to the zero power always gives a value of one.

The exact same approach is used in the binary numbering system. Whenever a value larger than 1 (the highest available digit) is needed, a new column is started to the left, representing the next higher power of two. For example:

$$\begin{aligned}
1011\ 0011 &= (1 \times 2^8) + (0 \times 2^7) + (1 \times 2^6) + (1 \times 2^5) + \\
&\quad (0 \times 2^4) + (0 \times 2^3) + (1 \times 2^1) + (1 \times 2^0) \\
&= (1 \times 128) + (0 \times 64) + (1 \times 32) + (1 \times 16) \\
&\quad + (0 \times 8) + (0 \times 4) + (1 \times 2) + (1 \times 1) \\
&= 128 + 0 + 32 + 16 + 0 + 0 + 2 + 1 \\
&= 179
\end{aligned}$$

Don't let all of this throw you. It doesn't really matter if you're not too good with math. In electronics work with digital circuitry, you will very rarely (if ever) be called upon to convert between the decimal and binary numbering systems. The important thing here is to have a rough idea of how the binary system works.

Bits are usually written in sets of four. This is done just as a matter of convenience. 1011 0011 is easier for a human being to read and copy without error than 10110011. The standard grouping of bits is the eight bit *byte*. The smaller group of four bits is known as a *nibble*. It really is. Who says electronic technicians don't have a sense of humor? Any larger grouping of bits, such as sixteen bits, or thirty-two bits, is usually called a *word*, or a *binary word*.

Digital circuitry offers numerous advantages over analog circuitry in many (but certainly not all) applications. Digital circuitry tends to be more immune to noise than analog circuitry. Input and output values are less ambiguous in a digital system. Advanced mathematical functions and programmability are easy to achieve with digital circuitry, but might be very difficult, or utterly impossible with analog circuitry.

In some applications the high or low nature of digital signals may be highly undesirable, so analog circuitry is still in widespread use, along with the newer digital circuitry. In many systems it is necessary to convert signals from a digital circuit for use in an analog circuit, or vice versa. Fortunately, this is not too difficult to do. To convert a digital signal into analog form, a circuit known as a *D/A converter* (digital/analog converter) is used.

132 Pulse circuits

To go in the opposite direction, and change an analog voltage into digital form, an *A/D converter* (analog/digital converter) is called for. These concepts will be used in the following projects.

SIMPLE D/A CONVERTER

D/A converter circuits are often built around operational amplifiers. One of the simplest designs is illustrated in Fig. 8-8. A suitable parts list for this project is given in Table 8-5.

Actually, this circuit should look fairly familiar to you. It's really nothing more than a simple summing amplifier like Project 1 in chapter 2. IC1A and its associated components form an inverting summer, and IC1B is an inverting voltage follower, to restore the original signal polarity.

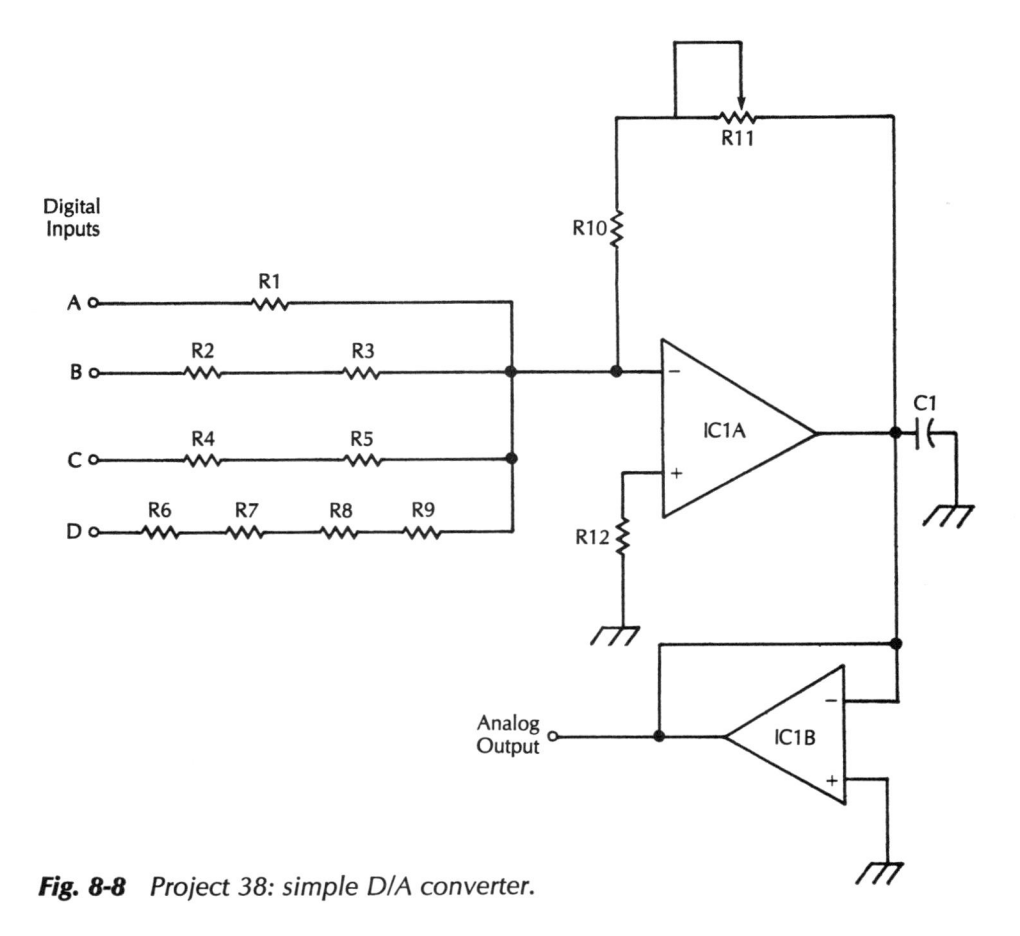

Fig. 8-8 *Project 38: simple D/A converter.*

Table 8-5. Parts List for Project 38: Simple D/A Converter.

Part	Description
IC1	747 dual op amp
R1, R2, R3	10K ¼-watt 5% resistor
R4, R6, R8	39K ¼-watt 5% resistor
R5, R7, R9	1K ¼-watt 5% resistor
R10	4.7K ¼-watt 5% resistor
R11	5K trimpot
R12	6.2K ¼-watt 5% resistor

The secret of using a summing amplifier as a D/A converter is in the weighted values of the input resistors. In a straight summing amplifier, all of the input resistors are normally given equal values, for unweighted inputs. All input signals are affected equally. In the D/A converter, however, each successive input resistor has exactly double the value of the preceding input resistor (double because you are converting a binary value). For best results, precision, low tolerance resistors should be used in this circuit. Alternately, you could use an ohmmeter to match the resistors as closely as possible.

As shown here, the D/A converter circuit accepts four bits (a nibble) in parallel. The most significant bit (MSB), or the leftmost bit in written form, is fed through resistance R. The second bit is fed through 2R, and the third bit passes through a resistance of 4R. Finally, the least significant bit (LSB), or rightmost bit, is fed through a resistance of 8R. Each successive bit experiences twice as much resistance as its predecessor, so they are unequally weighted. The most significant bit (MSB) passes through the smallest input resistance, resulting in the strongest signal applied to the op amp. Similarly, the least significant bit (LSB) passes through the largest input, giving the smallest signal, as seen by the op amp.

Remember that the inputs are all digital bits. They are either low (0 volts), or high (V+), with no intermediate values. A 0 input bit is merely ignored by the circuit. A 1 (or high) input bit passes through the appropriate weighted resistance, so that the op amp sees a different, discrete voltage for each bit input.

134 Pulse circuits

In this circuit R is set at 10K (10,000 ohms). This means, the bits, in order, are affected by the following resistances:

Bit			
1	MSB	R	10K
2		2R	20K
3		4R	40K
4	LSB	8R	80K

Notice that these are not all standard resistor values. For 20K, use a series combination of two 10K resistors. A 1K resistor in series with a 39K resistor gives you 40K, and two such pairs in series create 80K.

Thanks to Ohm's Law, the voltages seen by the op amp are proportional to the input resistances. For example, if bit 4 (LSB) is 1 volt, then bit 3 will be 2 volts, bit 2 will be 4 volts, and bit 1 (MSB) will be 8 volts. The output voltage will therefore be directly proportional to the binary value of the four input bits. Consider a couple of examples:

$$1010 = 8 + 0 + 2 + 0$$
$$= 10 \text{ volts}$$
$$0111 = 0 + 4 + 2 + 1$$
$$= 7 \text{ volts}$$
$$1111 = 8 + 4 + 2 + 1$$
$$= 15 \text{ volts (maximum possible value)}$$

In a practical circuit, the input voltages will probably be lower than this, for two reasons. One, most digital circuits put out less than 8 volts as the high signal, and two, reduced voltages are used to avoid overloading the input of the op amp. The op amp's supply voltage must be higher than the maximum voltage ever applied to its input, or clipping (and possible circuit damage) will occur.

Trimpot R11 is adjusted to set the overall gain of the amplifier. All of the inputs see an equal amount of gain, once they have passed through their (fixed) weighting input resistors. For some applications, you may decide to omit the gain control altogether, and just use an appropriate fixed resistor in the op amp's feedback loop.

The capacitor (C1) acts as a simple low-pass filter. It helps smooth out the output signal with changing input values. In some applications, this filter capacitor may be omitted. In other applications, a larger capacitor, or even a more sophisticated low-pass filter circuit may be desirable.

In theory, this D/A converter circuit can be expanded for more than four input bits. In practice, however, the amount of available expansion is severely limited. Since each additional input resistor must have double the preceding resistance value, the values quickly get awkward. Larger resistances are likely to be less precise and less stable. The weighted input voltages will become increasingly small, and the circuit will become more prone to errors due to noise. The circuit can be expanded, but it is a risky procedure, at best.

R–2R D/A CONVERTER

One way to get around some of the problems of the simple D/A converter circuit of the last project is to use a R–2R D/A converter circuit, like the one shown in Fig. 8-9. A suitable parts list for this project is given in Table 8-6. Once again, IC1A and its associated components is a modified inverted summing amplifier, performing the actual digital to analog conversion, and IC1B is an inverting voltage follower, used to restore the original signal polarity at the output.

The big advantage of this design is that only two basic resistor values are required for all of the inputs: R and 2R. Two R resistors can be used in series to create a 2R resistor. This is the approach suggested in the schematic and the parts list. In the four bit version of the circuit shown here, the following resistor equalities are used to create the R–2R ladder:

$$R_4 = R_6 = R_8 = R_{11} = R = 10K$$
$$R_1 = R_2 = R_5 = R_7 = R_9 = R_{10} = 2R = 20K$$

Four binary inputs are used in this circuit, but it can easily be expanded to include almost any number of input bits. However, if a very large number of input bits are used, there will still be the problem of extremely low voltages for the least significant bits.

R3 is shown as a potentiometer here. This permits manual adjustment of the amplifier's overall gain. All of the inputs will be equally affected by the gain control. In some applications, this may be a trimpot, set only during calibration, then left alone. In other applications, a fixed resistor of an appropriate value might be more suitable in the feedback loop of the op amp.

136 Pulse circuits

Fig. 8-9 Project 39: R-2R D/A converter.

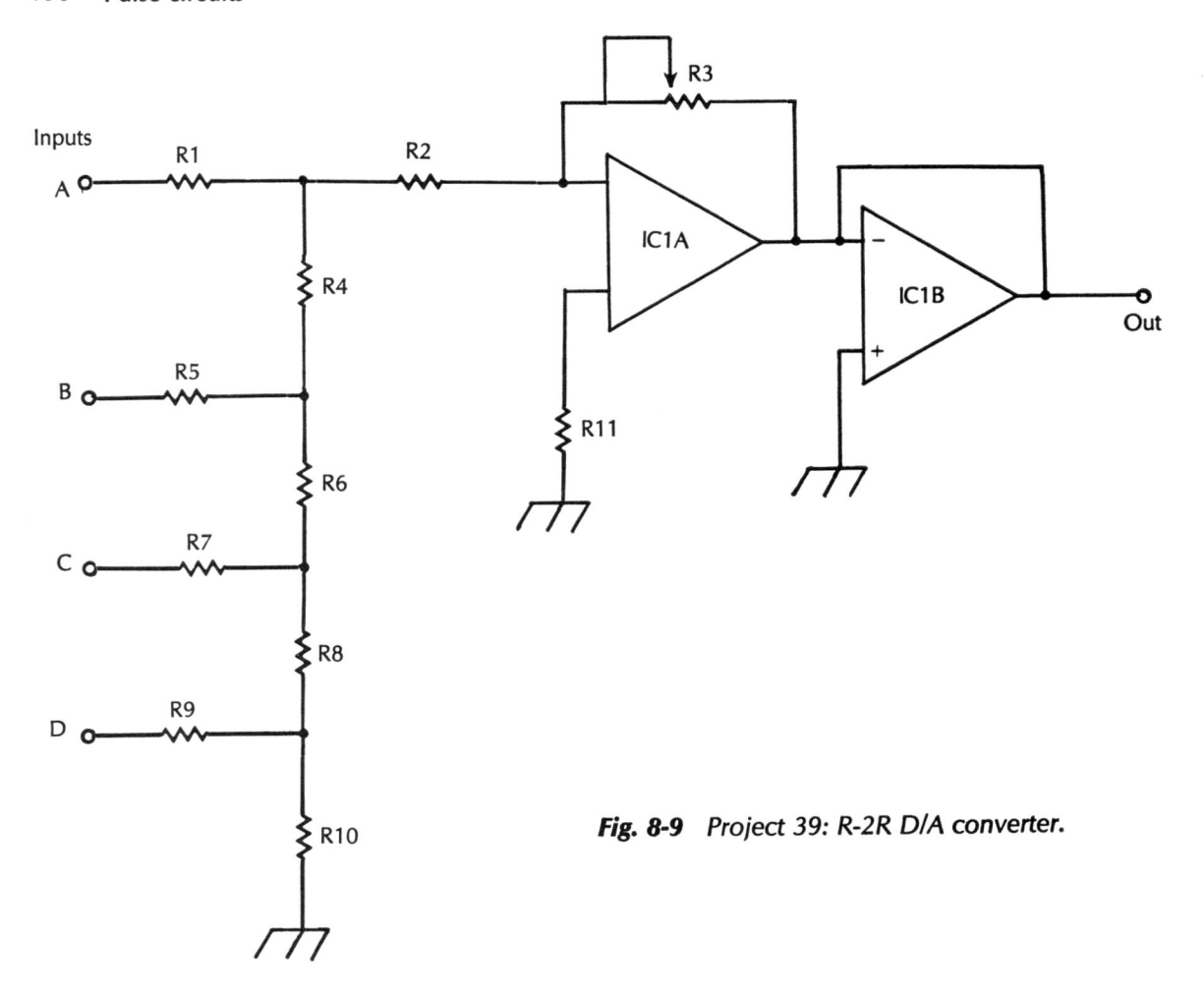

Table 8-6. Parts List for
Project 39: R-2R D/A Converter.

Part	Description
IC1	747 dual op amp
R1, R2, R5, R7, R9, R10	20K 1/4-watt 5% resistor (see text)
R4, R6, R8, R11	10K 1/4-watt 5% resistor
R3	50K potentiometer

The R-2R D/A converter circuit works because lower order bits see a greater resistance than higher order bits. If you follow the current paths in the schematic diagram, you can see that bit A (the most significant bit, or MSB) passes through only resistors R1 and R2, for a total resistance of:

$$R_a = R_1 + R_2$$
$$= 2R + 2R$$
$$= 4R$$

The second input bit (B) must pass through three resistors—R5, R4, and R2:

$$R_b = R_5 + R_4 + R_2$$
$$= 2R + R + 2R$$
$$= 5R$$

For bit C (the third input bit), four resistors appear in the current path: R7, R6, R4, and R2:

$$R_c = R_7 + R_6 + R_4 + R_2$$
$$= 2R + R + R + 2R$$
$$= 6R$$

Finally, the least significant bit (LSB), or bit D must pass through five resistors: R9, R8, R6, R4, and R2:

$$R_d = R_9 + R_8 + R_6 + R_4 + R_2$$
$$= 2R + R + R + R + 2R$$
$$= 7R$$

At the same time, the higher the order of the bit, the greater the number of resistors between the initial input resistor (R1, R5, R7, or R9) and ground. This creates a voltage divider. Bit A (MSB) has four resistors to ground (R4, R6, R8, and R10), while bit D (LSB) has just one (R10).

As a result of these varying resistances, each input bit voltage is properly weighted by the time it reaches the op amp. Each bit's input position in the R-2R ladder network forms a weighted voltage divider between the op amp's inverting input and ground.

IMPROVED D/A CONVERTER

A more sophisticated D/A converter circuit is illustrated in Fig. 8-12. A suitable parts list for this project is given in Table 8-7.

138 Pulse circuits

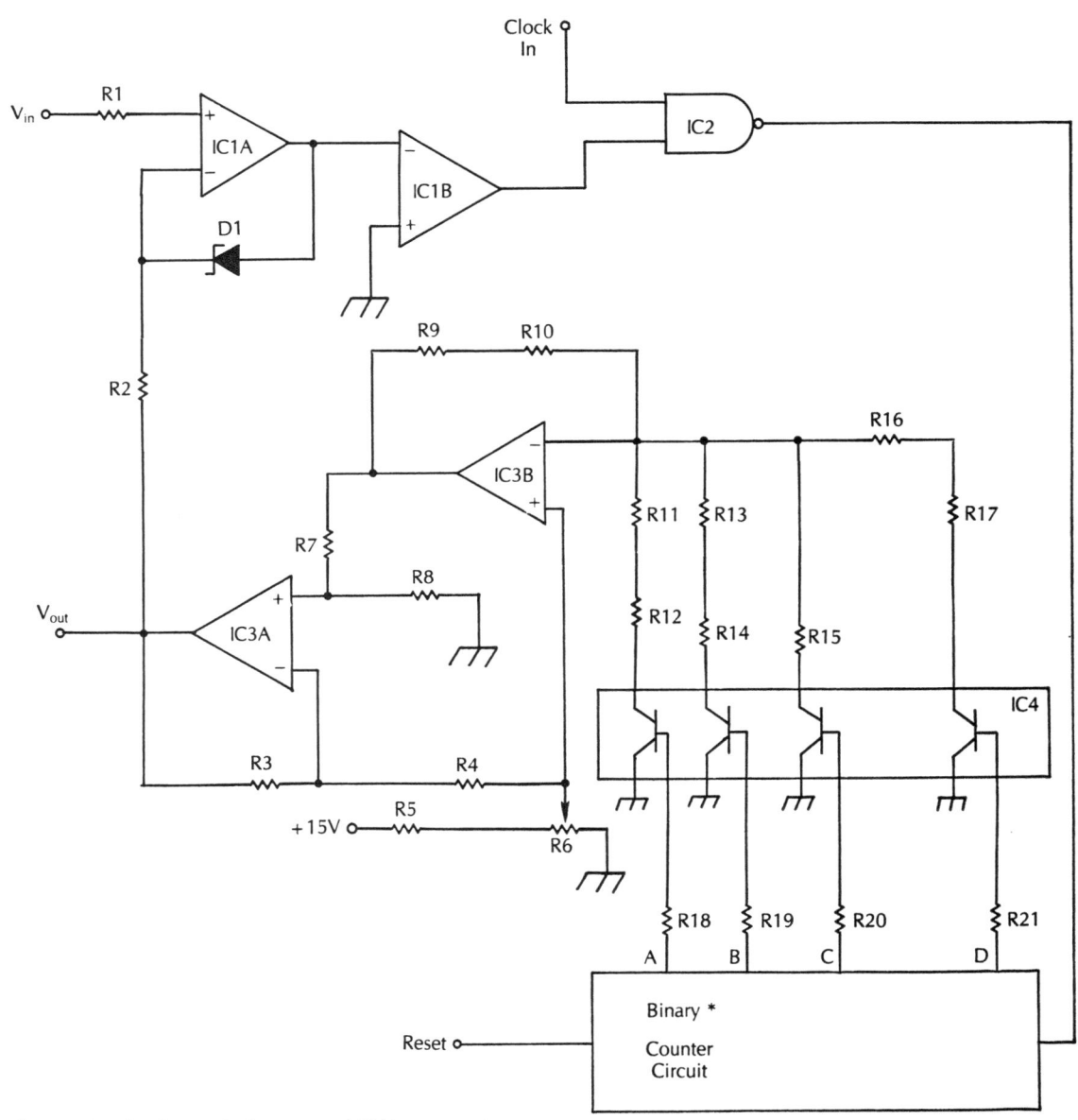

Fig. 8-10 *Project 40: improved D/A converter.*   * Not on Parts List—See Text.

Notice that this project is not shown fully complete here. Any four bit binary counter circuit can be used to complete the D/A converter. Any other four bit digital signal source may be substituted. A straightforward counter can be used for demonstration purposes. This portion of the circuit is shown in block

Table 8-7. Parts List for
Project 40: Improved D/A Converter.

Part	Description
IC1, IC3	747 dual op amp
IC2	CD4011 quad NAND gate
IC4	CA3026 transistor array (or similar)
D1	zener diode (select for X output voltage)
R1 – R4, R7, R8	10K 1/4-watt 5% resistor
R5	1.8K 1/4-watt 5% resistor
R6	1K trimpot
R9, R10	100K 1/4-watt 5% resistor
R11	220K 1/4-watt 5% resistor
R12, R14	33K 1/4-watt 5% resistor
R13	470K 1/4-watt 5% resistor
R15, R16, R17	1M 1/4-watt 5% resistor
R18, R19, R20, R21	2.7K 1/4-watt 5% resistor

diagram form here to simplify the schematic. The binary counter activates each transistor of IC4 through a base resistor (R18 through R21). When activated, each transistor effectively clamps the appropriate resistor(s) to ground.

For the best results substitute precision resistors as follows:

R11 – R12	250K precision resistor
R13 – R14	500K precision resistor
R15	1 Megohm precision resistor
R16 – R17	2 Megohm precision resistor

(Each resistor pair, such as R11 and R12, is replaced with a single precision resistor of the stated value.) For less critical applications the 5% resistors called for in the parts list will do reasonably well. The selected resistor (or resistors) set the amplifier gain.

The unusual feature of this circuit is that the binary bits control the amplifier gain for a fixed reference voltage input (V_{in}), by determining which transistors are switched on. A high (1) bit will turn on the appropriate transistor, while a low (0) bit will leave that transistor in the off state, and its resistor(s) switched out of the gain circuit. The output voltage from this circuit will always be positive, ranging from zero volts up to 10 volts, depending on the binary value of the input.

❖ 9
Miscellaneous projects

THIS FINAL CHAPTER IS A COLLECTION OF MISCELLANEOUS DUAL OP amp circuits which don't quite fit into any of the other chapter headings, but are still worthwhile and interesting.

DUAL-POLARITY VOLTAGE REGULATOR

Op amps usually aren't thought of in the area of power supply circuits, but these surprisingly versatile chips can be useful here, too. Many electronic circuits (including many op amp circuits) require a dual polarity power supply. That is, the power supply must provide two voltages: one positive with respect to ground, and the other negative with respect to ground. Generally speaking, these two voltages are equal, except for their opposite polarities.

Any critical circuit usually demands a regulated voltage as its power source, to prevent voltage fluctuations as the amount of current drawn by the circuit changes, and to limit the effects of noise spikes in the power supply line. A dual polarity op amp voltage regulator circuit is shown in Fig. 9-1. A suitable parts list for this project appears as Table 9-1.

In this circuit, a dual polarity unregulated voltage source serves as the input. The output is a regulated dual-polarity voltage source referenced to true earth ground. To keep things simple, split this circuit up into its positive and negative sections. First examine the positive portion of the regulator circuit. This section of the circuit is comprised of IC1A, Q1, D1, R1, R2, and R3. For convenience, this part of the circuit is shown separately in Fig. 9-2.

142 Miscellaneous projects

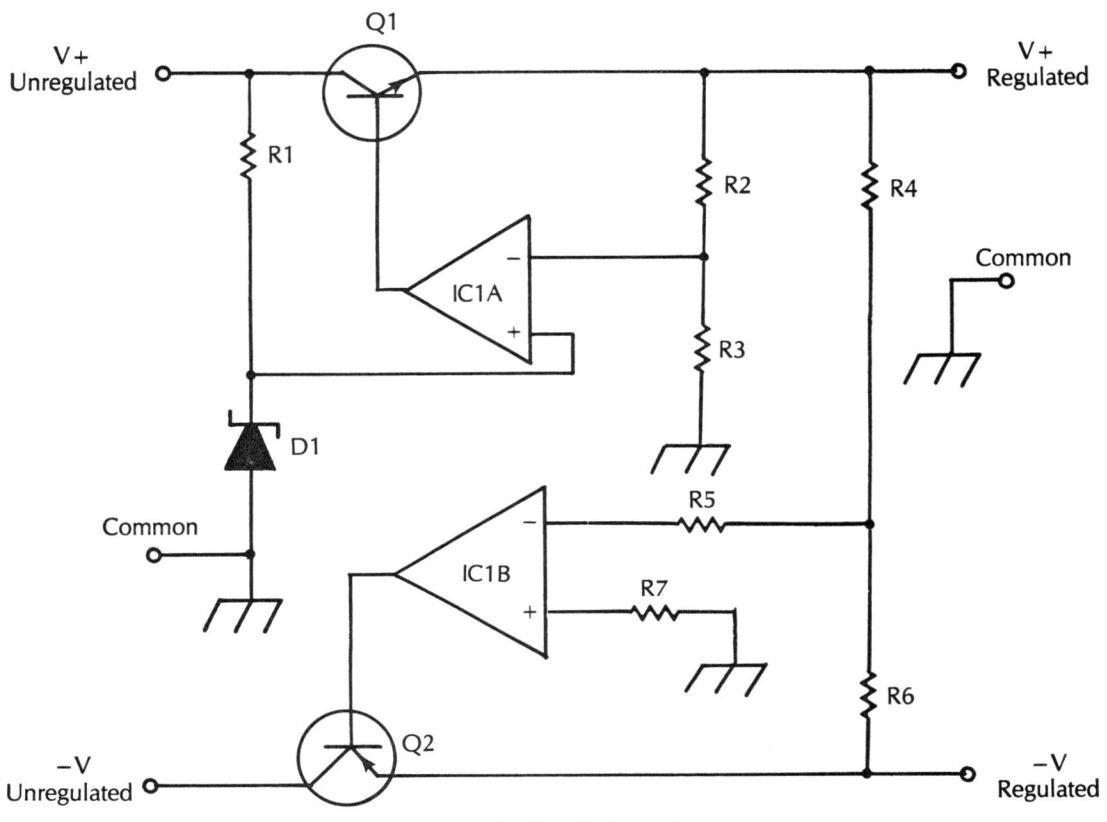

Fig. 9-1 Project 41: dual polarity voltage regulator.

Table 9-1. Parts List for Project 41: Dual Polarity Voltage Regulator.

Part	Description
IC1	747 dual op amp
Q1	npn transistor
Q2	pnp transistor
D1	9.1-volt zener diode
R1, R4, R6	12K 5% 1/2-watt resistor
R2, R3	2.2K 5% 1/2-watt resistor
R5, R7	10K 5% 1/2-watt resistor

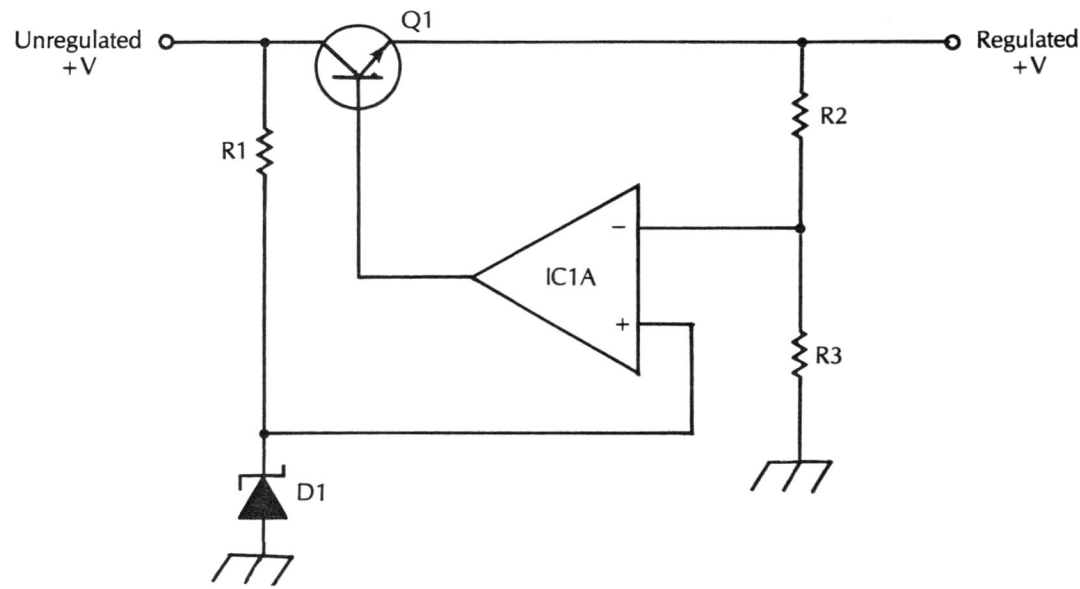

Fig. 9-2 *The positive voltage regulator portion of the dual polarity voltage regulator circuit of Fig. 9-1.*

Transistor Q1 is an npn series pass element. Its characteristics and specifications aren't too important in this particular application, as long as the transistor used can safely handle the maximum current drawn by the load. The base current of transistor Q1 is controlled directly by the output of the op amp (IC1A). The noninverting input signal to this op amp is a reference voltage determined by resistor R1 and zener diode D1. A portion of the op amp's voltage is fed back into the inverting input through a voltage divider network made up of resistors R2 and R3.

If the inverting input is fed the same voltage as the noninverting input, the operational amplifier will be in a stable condition, holding the output voltage constant. If the output voltage should attempt to increase or decrease for any reason, the voltage seen at the inverting input will be changed proportionately. This voltage will no longer be equal to the reference voltage set by R1 and D1. This causes the op amp's output signal to change, altering the base current of the transistor. The transistor's emitter voltage (the output) will be quickly brought back into line with the nominal output voltage. Even a very small change in the output

144 Miscellaneous projects

voltage will cause the circuit to respond very quickly to self-correct the error, making output ripple negligible and virtually nonexistent.

Now, what about the negative voltage half of the circuit? You could just construct a mirror image of the positive voltage regulator circuitry, and reverse all the polarities. This would not be a very elegant or interesting solution. To reduce the parts count of the project, and to ensure that the dual-polarity outputs will be truly symmetrical around earth ground, a slightly different approach is employed in this section of the circuit.

Figure 9-3 shows just the negative portion of the dual-polarity op amp voltage regulator circuit. This section of the circuitry includes IC1B, Q2, R4, R5, R6, and R7. Notice that no zener diode is needed to set up a reference voltage in this part of the circuit. The positive regulated output voltage is used as the reference voltage source.

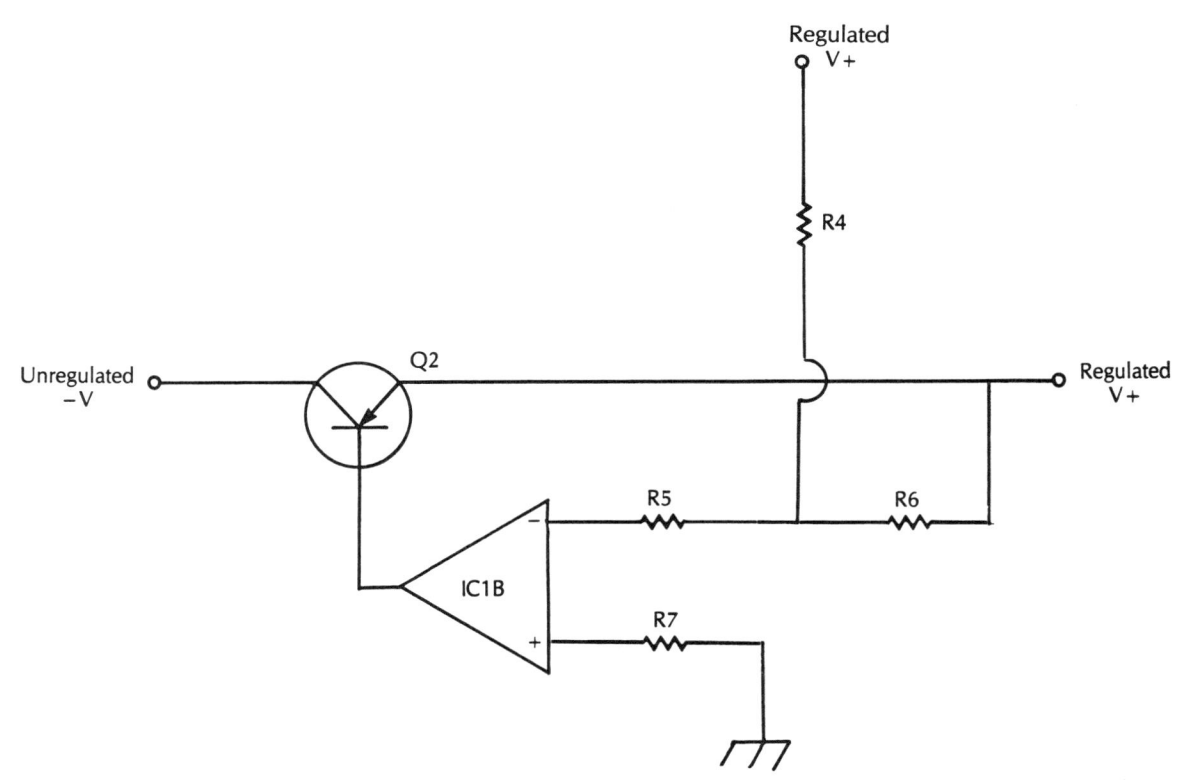

Fig. 9-3 *The negative voltage regulator portion of the dual polarity voltage regulator circuit of Fig. 9-1.*

Transistor Q2 is a pnp unit, to account for the change of polarity in this portion of the circuit. Once again, the characteristics and specifications for this transistor aren't too critical in this circuit, as long as the transistor used can safely handle the maximum current drawn by the load. For proper symmetry, pnp transistor Q2 should be selected to match npn transistor Q1 as closely as possible.

Resistor R7 references the noninverting input of the op amp (IC1B) to ground. Resistors R4 and R6 have identical values. For very critical applications, it's a good idea to use precision resistors for R4 and R6, or select closely matched units with an ohmmeter. Because these two resistors have equal values, the voltage at their junction will be zero if the regulated +V output voltage (reference) is equal to the regulated −V output voltage. This junction signal is fed into the op amp's inverting input through resistor R5.

If the two regulated output voltages are not exactly equal, a nonzero signal will be fed into the inverting input of IC1B, causing the op amp's output voltage to shift, altering the base current of transistor Q2. Since the +V voltage is regulated by the other half of the circuit, any changes in Q2's base current cause it to adjust the −V output voltage.

One major advantage of this dual polarity voltage regulator circuit is that even if the output voltages change momentarily, the outputs will still maintain symmetry, since the −V regulated output voltage is referenced directly to the +V regulated output voltage. This characteristic can be of major importance in many precision op amp circuits.

PRECISION FULL-WAVE RECTIFIER

Operational amplifiers are sometimes employed as precision diodes, especially in small signal applications. An ordinary semiconductor diode (or rectifier) has a certain switching voltage, due to the voltage drop across the device. For silicon diodes, this voltage drop is approximately 0.7 volt. In germanium diodes, the voltage drop is about 0.3 volt.

If a signal less than the voltage drop is placed across a diode, the signal will be blocked, even though the device is forward biased. When rectifying a low voltage ac signal, a portion of the input wave is lost, as illustrated in Fig. 9-4. The lower the signal

146 Miscellaneous projects

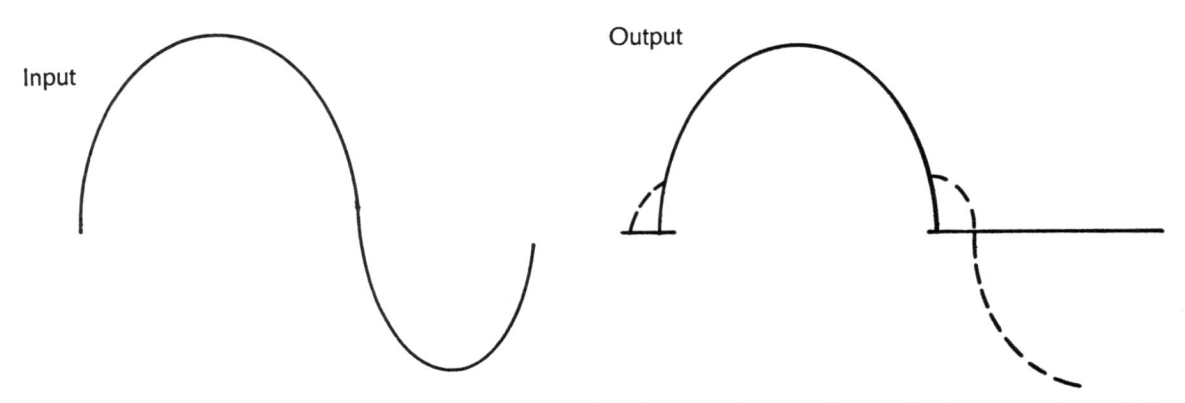

Fig. 9-4 *When rectifying a low-voltage ac signal with an ordinary silicon diode, a portion of the input waveform is lost.*

amplitude, the more significant this effect will be. In a low signal rectification application, an ordinary semiconductor diode just doesn't do a very good job.

This is where a precision diode circuit comes in handy. In effect, the circuit behaves like an ordinary diode, except it does not have the low signal problem discussed above. The voltage drop across a precision diode is essentially zero. (Actually, it is very, very small, and can reasonably be ignored in most practical applications.) Figure 9-5 shows a precision full-wave rectifier cir-

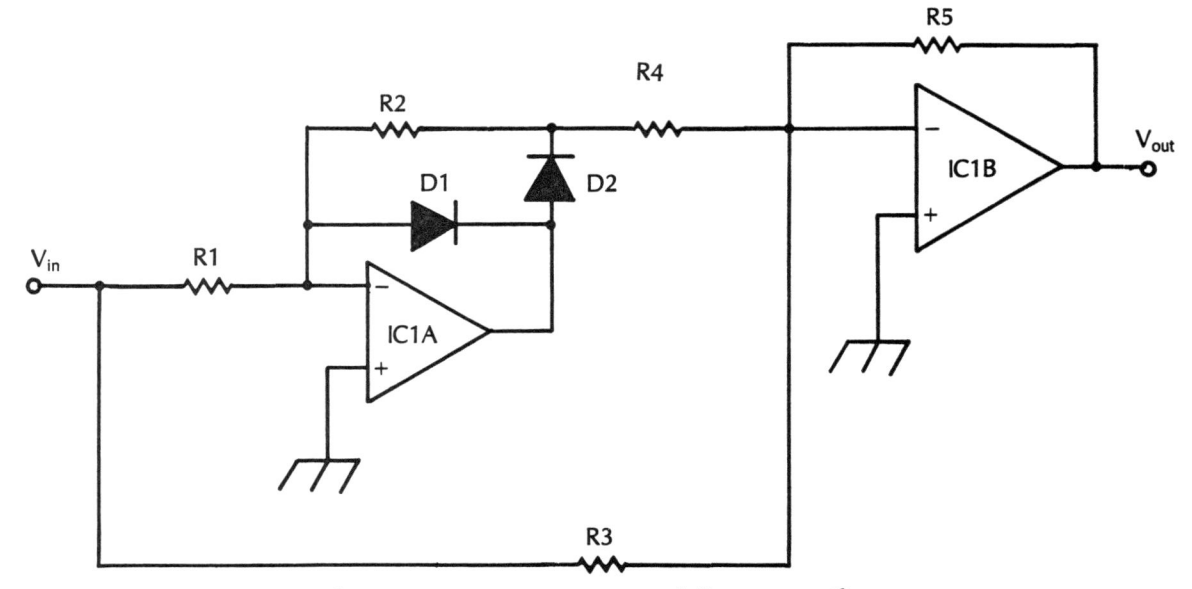

Fig. 9-5 *Project 42: precision full-wave rectifier.*

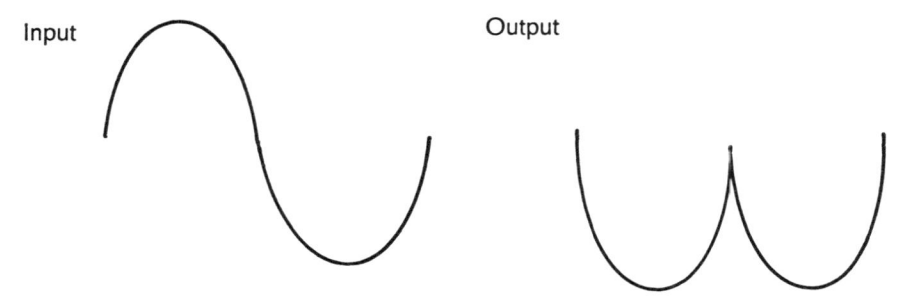

Fig. 9-6 *The precision full-wave rectifier circuit of Fig. 9-5 passes both half-cycles of the input ac waveform.*

Table 9-2. Parts List for Project 42:
Precision Full-Wave Rectifier.

Part	Description
IC1	747 dual op amp
D1, D2	diode (1N4001, or similar)
R1, R2, R4	10K ½-watt 5% resistor
R3, R5	22K ½-watt 5% resistor

cuit. It passes both halves of the ac cycle at the input, as illustrated in Fig. 9-6. A suitable parts list for this project is given in Table 9-2.

IC1A and its associated components form the actual precision diode in this circuit. For the time being, ignore IC1B, and resistors R3, R4, and R5. This leaves just a precision diode, or half-wave rectifier, as shown in Fig. 9-7.

When the input signal is positive, diode D1 is forward biased, and diode D2 is reverse biased. All of the feedback current flows through D1, so the voltage at the output is zero. The situation is reversed when the input voltage is negative. Now current flows through feedback diode D2, but not through diode D1. An inverted version of the input voltage now appears at the output.

In the full-wave rectifier circuit of Fig. 9-5, the precision diode network just discussed, is combined with a summing amplifier (IC1B, and its associated resistors). The input and output voltages of the half-wave rectifier are summed together. When the input has positive half-cycles, the output of the precision

148 Miscellaneous projects

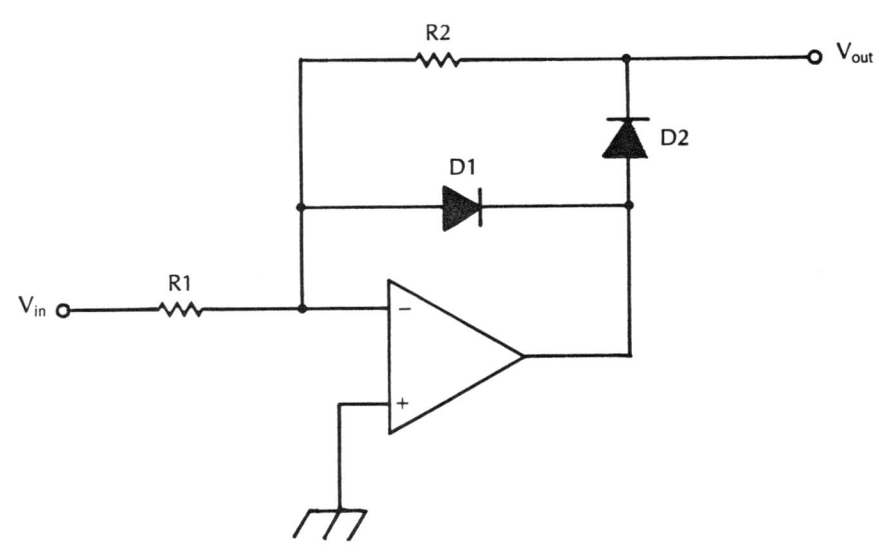

Fig. 9-7 *This precision diode network is used in the precision full-wave rectifier circuit of Fig. 9-5.*

diode (IC1A) is zero. This portion of the input waveform is simply amplified and inverted by IC1B. Because the input resistor (R3) and the feedback resistor (R5) have equal values, this stage operates at unity gain. The output voltage equals the input voltage, except for the polarity inversion.

When the input signal is in its negative half-cycle, the precision diode conducts. Its output is positive because of the signal inversion performed by the op amp in this stage. So, there are positive half-cycles (from the precision diode) being fed into the summing amplifier, along with the negative half-cycles from the original input signal. The input resistors for these two signals are weighted. Resistor R4 (precision diode signal) has one-half the value of resistor R3 (input signal). Half the precision diode signal is canceled by the original input signal. The other half is amplified by inverting amplifier IC1B. In this case, the appropriate input resistor (R4) has half the value of the feedback resistor (R5), so the amplifier's gain is 2. The signal is doubled back to its original level. The output is a negative half-cycle.

To restate the action, on positive half-cycles at the input, the output of this circuit is a negative half-cycle, and when the input signal is going through its negative half-cycle, the circuit's output will be another negative half-cycle. The circuit performs full wave rectification, outputting a negative pulsating dc signal. If your specific application needs the rectified signal to be positive rather than negative, just pass the signal through an inverting voltage follower stage.

For maximum precision, resistors R3 and R5 should be 20K, rather than the standard 22K resistors specified in the parts list. You could use a precision resistor, or two 10K resistors in series for each of these 20K resistors.

HIGH INPUT IMPEDANCE INVERTING AMPLIFIER

Operational amplifiers have pretty high input impedances. Typical values run from about 300K (300,000) to thousands of Megohms (1 Megohm = 1,000,000 ohms). This is desirable, because the high input impedance minimizes loading of the signal source, or preceding stage in the circuit.

In practical op amp circuits, however, the actual effective input impedance is significantly less than that of the op amp device itself. For example, in a standard inverting amplifier circuit, the actual input impedance seen by the load is equal to the resistance of the circuit's input resistor. This component may have a value as low as 10K (10,000 ohms), or even 1K (1,000 ohms). As a result, loading of the source circuitry can be a real problem in many practical applications.

One solution is to use the circuit shown in Fig. 9-8. A typical parts list for this project is given in Table 9-3. This circuit uses a technique known as controlled positive feedback, or *bootstrapping*, to boost the effective input impedance of the circuit. The greater the positive feedback, the higher the input impedance. There is a limit to how far this technique can be taken. If excessive positive feedback is used, the circuit will break into oscillation, which obviously is undesirable in an amplifier circuit.

In an inverting amplifier circuit, a second op amp stage is used to bootstrap the input impedance of the circuit. In this circuit, op amp IC1A is the actual inverting amplifier. Op amp IC1B supplies most, if not all of IC1A's input current through resistor R2. Without IC1B, the input impedance to IC1A would simply be equal to the value of resistor R1. With the bootstrapping technique, however, the effective input impedance is (ideally) equal to:

$$R_{in} = \frac{R1 \times R2}{R2 - R1}$$

150 Miscellaneous projects

Fig. 9-8 Project 43: high-input impedance inverting amplifier.

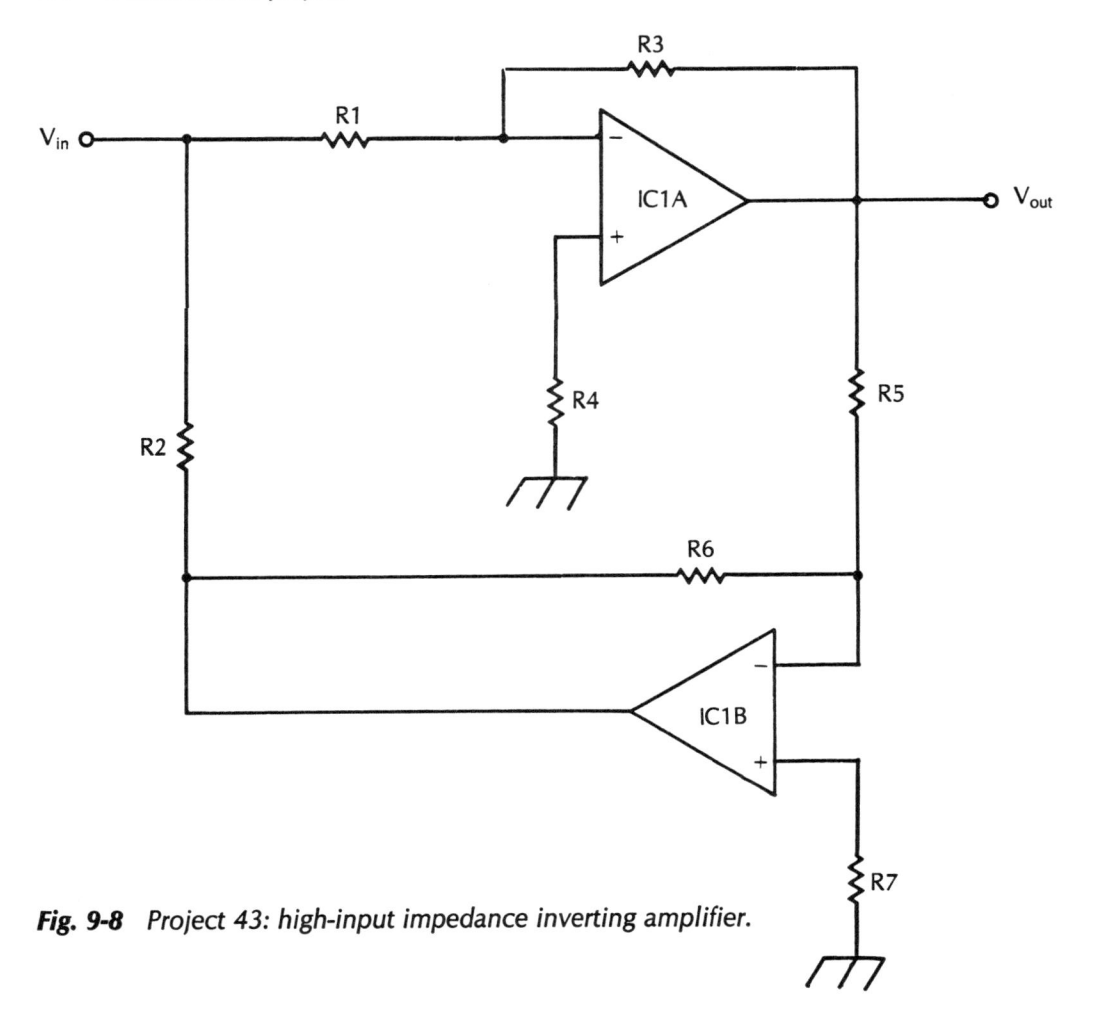

Table 9-3. Parts List for Project 43: High-Input Impedance Inverting Amplifier.

Part	Description
IC1	747 dual op amp
R1, R7	100K 1/4-watt 5% resistor
R2	82K 1/4-watt 5% resistor
R3, R5	470K 1/4-watt 5% resistor
R4	270K 1/4-watt 5% resistor
R6	220K 1/4-watt 5% resistor

For best results, and to avoid stability (oscillation) problems, the value of resistor R2 should be slightly less than that of resistor R1. According to the parts list, the following resistor values are recommended:

$$R1 = 100K \quad (100{,}000 \text{ ohms})$$
$$R2 = 82K \quad (82{,}000 \text{ ohms})$$

Using these resistor values, the effective input resistance for the circuit works out to:

$$R_{in} = \frac{100{,}000 \times 82{,}000}{82{,}000 - 100{,}000}$$
$$= \frac{8{,}200{,}000{,}000}{-18{,}000}$$
$$= -455{,}555 \text{ ohms}$$

You have achieved an increase of almost five times the original (nonbootstrapped) effective input impedance. Don't worry about the negative sign in the results of this equation. It has some significance to the phase of the signal but you can reasonably ignore it in most applications. A negative impedance is equivalent to a positive impedance of the same value. With ac signals there will be some phase shift at certain frequencies. This usually won't be much of a problem in most inverting amplifier applications.

The gain of the inverting amplifier itself, is simply the inverted feedback resistance (R_3), divided by the (true) input resistance (R_1):

$$G = -R_3/R_1$$

Using the suggested component values from the parts list, the gain works out to

$$G = -470{,}000/100{,}000$$
$$= -4.7$$

Only the input impedance is affected by the bootstrapping technique. The operation of the inverting amplifier circuit itself remains unchanged. For this bootstrapping circuit to function properly, the value of resistor R5 should be equal to the inverting amplifier's feedback resistor (R3), and resistor R6 should have double the value of the input resistor (R1).

AMPLIFIER WITH DIFFERENTIAL INPUTS AND OUTPUTS

In its most basic form, an operational amplifier has a single output and a pair of differential inputs. The output signal is equal to the difference between the two input signals, multiplied by the amplifier gain. The amplifier circuit shown in Fig. 9-9 not only has differential inputs, it also has differential outputs. A suitable parts list for this project appears as Table 9-4.

In some technical literature, a differential output, such as shown here, is sometimes referred to as a *double-ended output*. Whatever the name, the purpose is to drive a floating load. A floating signal is one which is not referenced directly to ground.

In an ordinary op amp circuit, the output signal is taken off between the op amp's actual output pin and circuit ground. In the differential input/output circuit of Fig. 9-9, neither the input

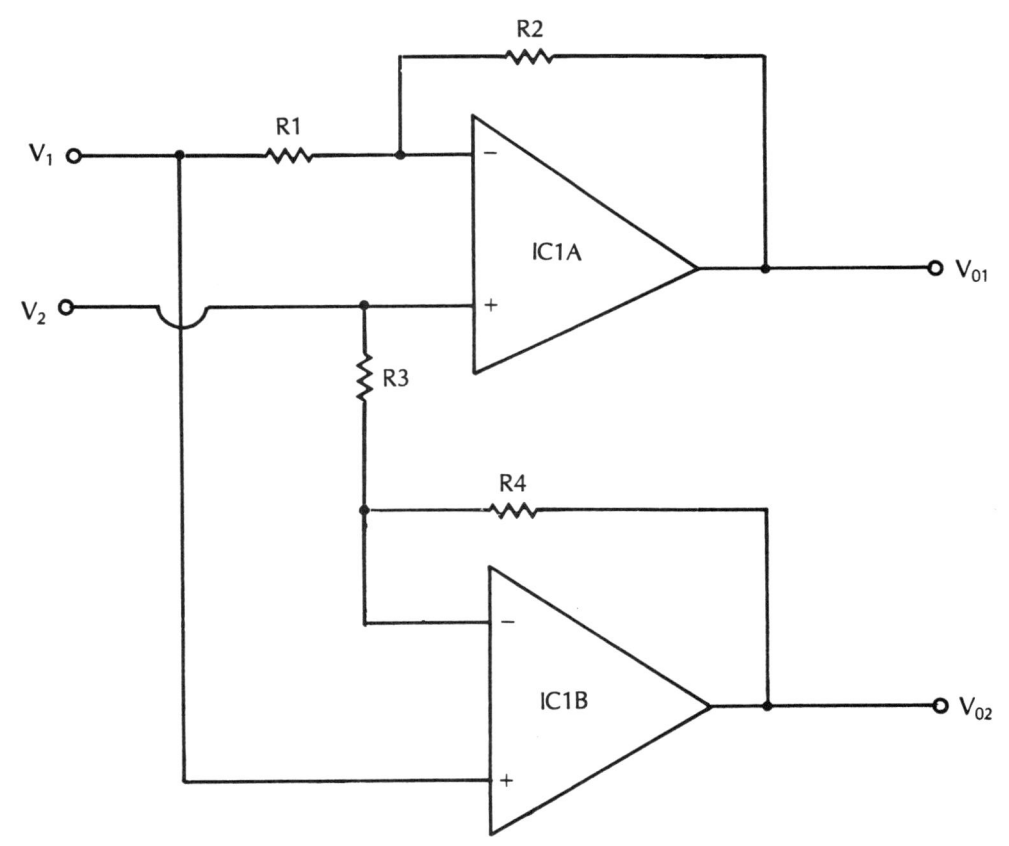

Fig. 9-9 Project 44: amplifier with differential inputs and outputs.

Amplifier with differential inputs and outputs 153

Table 9-4. Parts List for Project 44: Amplifier with Differential Inputs and Outputs.

Part	Description
IC1	747 dual op amp
R1, R3	10K 1/4-watt 5% resistor
R2, R4	100K 1/4-watt 5% resistor

signal, nor the output signal is referenced to ground. The output signal is tapped off from the differential outputs. There is no circuit ground connection in the signal line. The differential output voltage is equal to time difference between the two output voltages:

$$V_{out} = V_{out1} - V_{out2}$$

The system works very much like the usual differential inputs of an ordinary operational amplifier circuit. In effect, V_{out2} is the inverting output and V_{out1} is the noninverting output.

The input voltage to this circuit is also taken differentially. It is equal to the difference between the two inputs:

$$V_{in} = V_1 - V_2$$

For the circuit as a whole, V_1 serves as the noninverting input and V_2 as the inverting input.

The gain of the circuit is simply the sum of the gains for the two component amplifier circuits. That is:

$$G = (R_2/R_1 + R_3/R_4)$$

In the suggested parts list, resistor R1 is made equal to resistor R3, and resistor R2 is given the same value as resistor R4. Thus, both halves of the circuit exhibit the same gain, so the inputs and outputs are unweighted. If unequal gains are used in the two halves of the amplifier circuit, the inputs and outputs will be weighted. That is, the differential output voltage will be shifted with respect to ground (true zero).

Using the component values suggested in the parts list, the circuit's overall gain works out to:

$$\begin{aligned} G &= (100{,}000/10{,}000) + (100{,}000/10{,}000) \\ &= 10 + 10 \\ &= 20 \end{aligned}$$

154 Miscellaneous projects

Any desired gain can be set up simply by selecting the appropriate resistor values. The differential output voltage (V_{out}) from this circuit is equal to the differential input voltage (V_{in}) multiplied by the circuit gain:

$$\begin{aligned} V_{out} &= V_{out1} - V_{out2} \\ &= G \times V_{in} \\ &= (R_2/R_1) + (R_3/R_4) \times (V_1 - V_2) \end{aligned}$$

This differential input/output amplifier circuit will exhibit the best performance when the two op amps used in the circuit are very closely matched. As a rule of thumb, the best possible match will be between circuits on the same chip, so a dual op amp IC, such as the 747 is ideal for an application of this type. When using a dual op amp IC in this circuit, the total differential dc output offset error will be extremely close to zero. Even in critical applications, no external offset adjustments will ever be necessary. In this case, the dc output offset error is equal for each of the two op amps (on the same chip), so in this configuration, they cancel each other out. This almost certainly would not happen with two separate op amp devices, even if high-precision units were used.

DEADSPACE CIRCUIT

If you sum together the outputs of two precision diodes, the result will be what is known as a deadspace circuit. The schematic diagram for this project appears in Fig. 9-10, with a suitable parts list given in Table 9-5.

In this circuit, IC1A and its associated components form one precision diode, while the second precision diode is made up of IC1B and its associated components. IC2 (and its associated components) is the summing amplifier, blending the outputs of the other two stages.

Take a moment here to refresh your memory about precision diode circuits. For more information, refer back to the discussion of the precision full-wave rectifier project presented earlier in this chapter. Remember, any ordinary semiconductor diode (or rectifier) has a small voltage drop across it, even when it is forward-biased. For germanium diodes, this voltage drop is approximately 0.3 volt. In silicon diodes, the voltage drop is a bit higher, about 0.7 volt. If a signal less than the voltage drop is placed

Deadspace circuit 155

Fig. 9-10 Project 45: deadspace circuit.

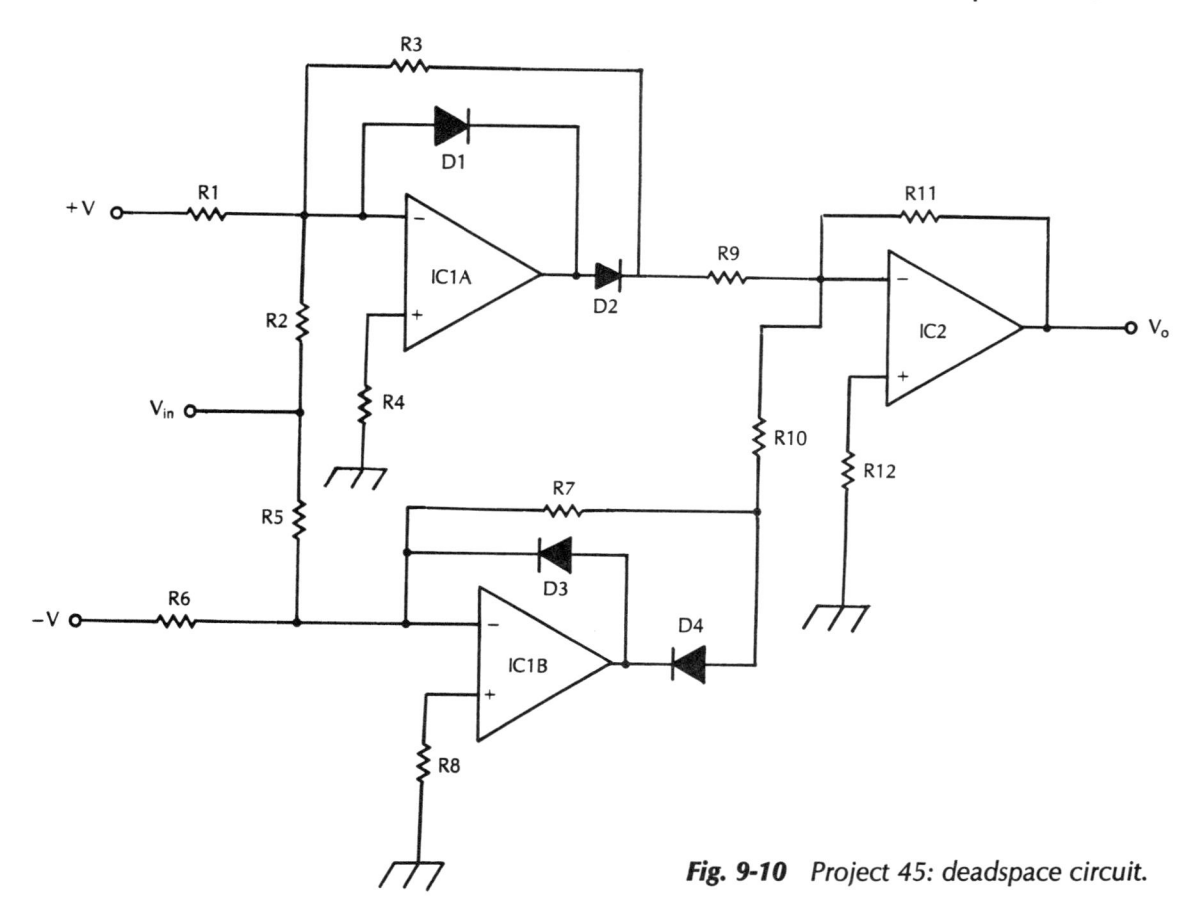

Table 9-5. Parts List for Project 45: Deadspace Circuit.

Part	Description
IC1	747 dual op amp
IC2	op amp (1/2 747, or 741, or similar)
D1 – D4	diode (1N4001, or similar)
R1	68K 1/4-watt 5% resistor
R2, R3, R5, R7, R9 – R11	10K 1/4-watt 5% resistor
R4	1.2K 1/4-watt 5% resistor
R6	33K 1/4-watt 5% resistor
R8	2.2K 1/4-watt 5% resistor
R12	4.7K 1/4-watt 5% resistor

across a diode, the signal will be blocked, even though the device is forward biased. When rectifying a low-voltage ac signal, a portion of the input wave is lost.

In many small signal rectification applications, a precision diode circuit will do a far better job than a standard semiconductor diode. In effect, the circuit behaves like an ordinary diode, except it does not have the low signal problem discussed above. Essentially, the voltage drop across a precision diode is zero. (Actually, it's very, very small, and can reasonably be ignored in most applications.)

For the moment, just look at the precision diode constructed around IC1A in the circuit of Fig. 9-10. What applies to this section of the circuit also applies to the circuitry around IC1B. When the input signal as seen by this portion of the circuit is positive, diode D1 is forward biased, and diode D2 is reverse biased. All of the feedback current flows through D1, so the voltage at the output is zero. Not surprisingly, the situation is reversed when the input voltage to this stage is negative. Now current flows through feedback diode D2, but not through diode D1. As inverted version of the input voltage now appears at the output.

The input signal (V_{in}) is combined with a positive reference voltage (equal to the circuit's positive supply voltage dropped through resistor R1) to form the input signal to the precision diode network constructed around IC1A. A negative reference voltage (equal to the circuit's negative supply voltage dropped through resistor R6) is combined with the input voltage (V_{in}) for the precision diode circuit built around IC1B.

These opposite reference voltages cause the two precision diode stages to see opposite polarity input signals. One of the precision diodes will be forward biased, while the other is reverse biased. But notice that the input resistors for the two reference voltages (R1 and R6) do not have equal values. This means that for certain low voltage input signals, both precision diode stages will be reverse biased, resulting in a 0-volt output.

For higher voltages (either positive, or negative), the circuit will function as an ordinary amplifier. The amplifier gain is set by the resistances in the summing amplifier stage—resistors R9, R10, and R11. But there will be a deadspace (zero output) in the middle of the circuit's operating range. This is illustrated in the typical output graph shown in Fig. 9-11.

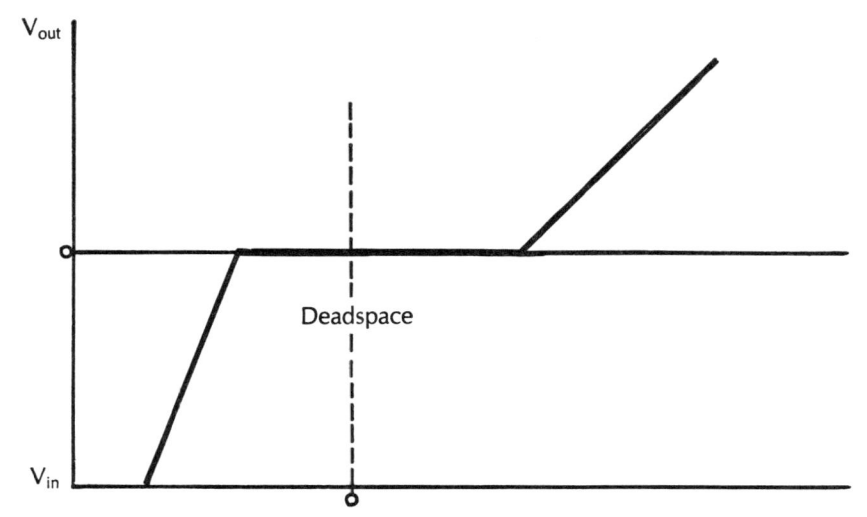

Fig. 9-11 *This graph illustrates the operation of the deadspace circuit of Fig. 9-10.*

The size of the deadspace will be determined by the relative values of the reference voltage input resistors (R1 and R6), and the input signal's input resistors (R2 and R5). Ordinarily, resistors R2 and R5 should have equal values. The values of resistors R1 and R6 should be significantly higher than the value used for R2 and R5.

Using the component value suggested in the parts list, the deadspace will run from about -2.5 volts to approximately $+5$ volts. If the reference voltage input resistors (R1 and R6) are given equal values, the deadspace will be symmetrical around zero. That is, it will extend an equal distance in both the positive and the negative directions.

Deadspace circuits, like this one, often are used in control systems to reduce the effects of noise in the control signal lines. The noise signal will typically be low enough to fall into the deadspace, and be eliminated by the circuit. The desired control signals are selected to be large enough to fall well outside the deadspace, and to be reliably passed through by this circuit.

SERIES LIMITER

Closely related to the deadspace circuit (Fig. 9-10) is the series limiter circuit. A practical circuit of this type is illustrated in Fig. 9-12. A suitable parts list for this project is given in Table

158 Miscellaneous projects

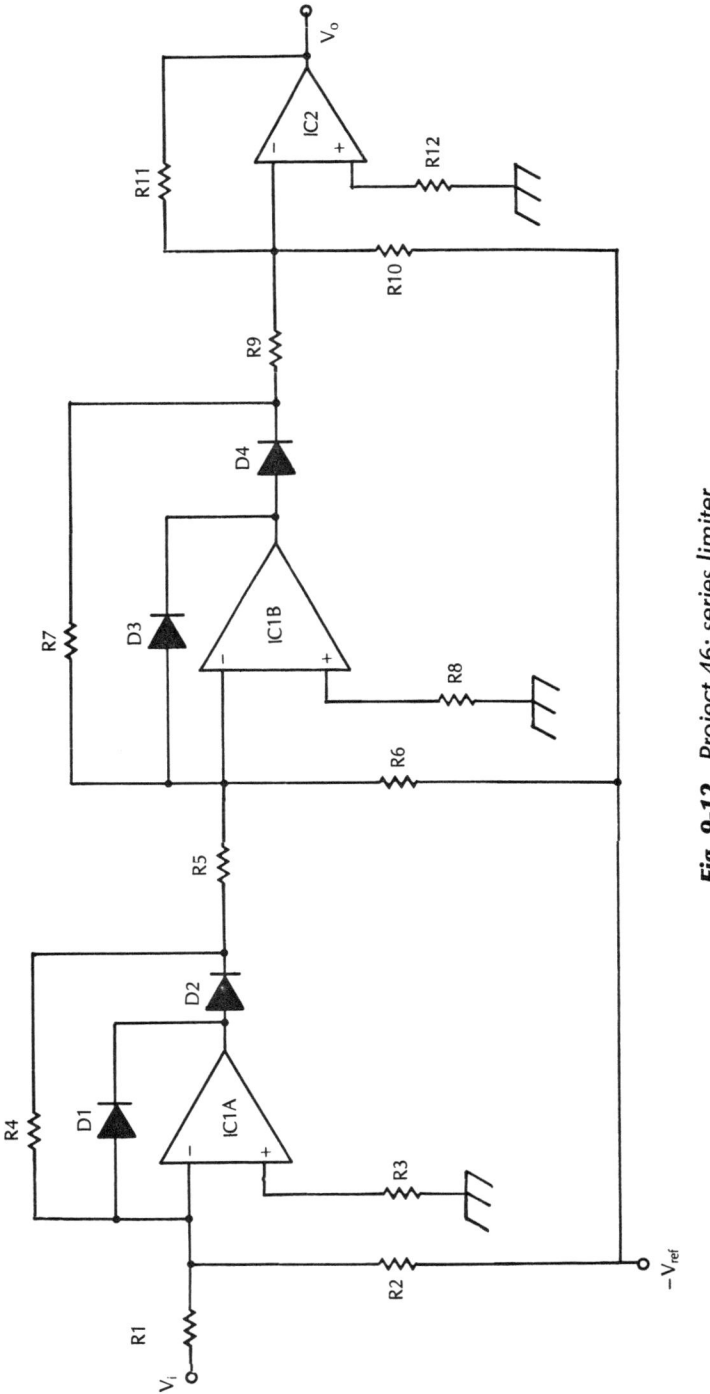

Fig. 9-12 *Project 46: series limiter.*

Table 9-6. Parts List for Project 46: Series Limiter.

Part	Description
IC1	747 dual op amp
IC2	op amp (½ 747, or 741, or similar)
D1–D4	diode (1N4001, or similar)
R1, R2, R4, R5, R7, R9–R11	10K ¼-watt 5% resistor
R3, R12	3.3K ¼-watt 5% resistor
R6	4.7K ¼-watt 5% resistor
R8	2.2K ¼-watt 5% resistor

9-6. The chief difference between the series limiter and the dead-space circuit is that in the series limiter, the two precision diode stages are in series, rather than in parallel.

Once again, the third stage in this circuit is a simple inverting summing amplifier (IC2, and its associated components). This third op amp may be one-half a second 747 chip, or a single op amp IC, such as the 741. IC1A and its associated components form one precision diode, while the second precision diode is made up of IC1B and its associated components.

A precision diode circuit is used when the normal voltage drop of an ordinary semiconductor diode (0.7 volt for silicon diodes, or 0.3 volt for germanium diodes) is undesirable, especially when rectifying small ac voltages. If a signal less than the voltage drop is placed across a standard semiconductor diode, the signal will be blocked, even though the device is forward biased.

In many small signal rectification applications, a precision diode circuit will do a far better job than a standard semiconductor diode. In effect, the precision diode circuit behaves much like an ordinary diode, except it does not have the low signal problem discussed above. Essentially, the voltage drop across a precision diode is zero. (Actually, it's very, very small, and can be reasonably ignored in most applications.)

For the moment, just look at the precision diode constructed around IC1A in the circuit of Fig. 9-12. What applies to this section of the circuit also applies to the circuitry around IC1B. Whenever the ac input signal to this portion of the circuit is in a positive half-cycle, diode D1 is forward biased, and diode D2 is reverse biased. All of the feedback current flows through D1, so the voltage at the output of the precision diode is zero. Not surprisingly, the situation is reversed when the input voltage to this stage goes negative. Now current flows through feedback diode D2, but not through diode D1. An inverted version of the input voltage now appears at the output.

In some ways, the series limiter circuit functions in just the opposite way as the deadspace circuit presented earlier in this chapter. In what was the deadspace region, the series limiter circuit behaves like an ordinary, linear amplifier. The output voltage is directly and linearly proportional to the input voltage. Outside this band, however, the output goes into a condition similar to saturation. The output voltage is clipped at the reference voltage level. If the input voltage is made more positive than the positive reference voltage, no further increases will be permitted in the output voltage. Similarly, making the input voltage more negative than the negative reference voltage will have no effect on the output voltage beyond the clipping point. The operation of this series limiter circuit is illustrated in Fig. 9-13.

In experimenting with alternate component values in this circuit, pay attention to the equal resistances in the suggested parts list. For the circuit to function properly, it is especially

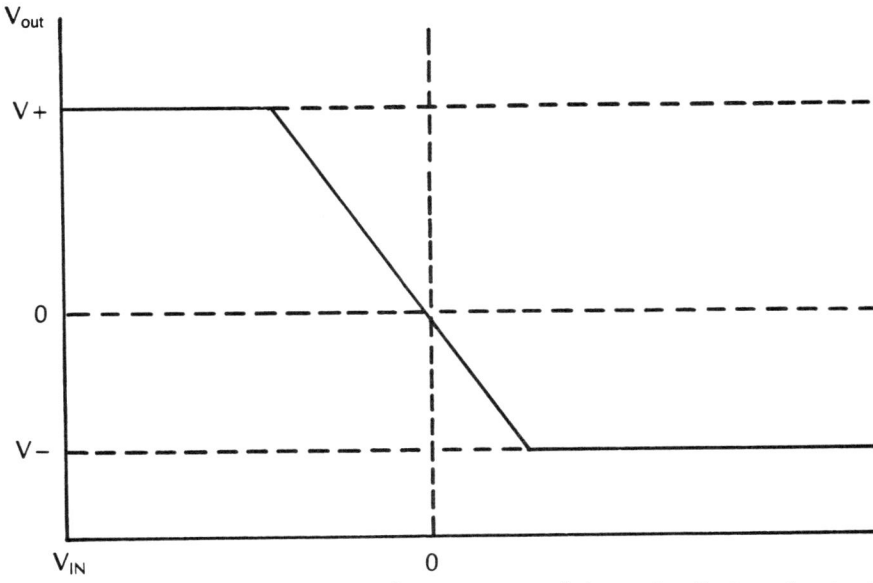

Fig. 9-13 *This graph illustrates the operation of the series limiter circuit of Fig. 9-12.*

important for the two input resistors (R1 and R2) to have equal values, matching the feedback resistances in the two precision diode stages. Resistor R8 should have about half this value.

LOW-IMPEDANCE INSTRUMENTATION AMPLIFIER

Operational amplifiers are widely used in *instrumentation amplifier* applications. An instrumentation amplifier is usually found in measurement applications. The term is not very precise, and is more one of convenience, than of exact definition. The remaining three projects will be of this type.

Most instrumentation amplifiers have certain common features. Generally speaking, the differential inputs of an operational amplifier will be used in an instrumentation amplifier. The output will usually be single-ended. Most instrumentation amplifier circuits are carefully designed for relatively high gain, with a minimum of output offset errors. A good instrumentation amplifier will normally have strong common mode rejection. That is, any signal common to both of the differential inputs will be canceled out and have no (or minimal) noticeable effect on the circuit's output signal. Usually, it is very desirable for an instrumentation amplifier circuit to have a very high input impedance, but in some specialized applications, a low input impedance may be preferred.

Figure 9-14 shows the schematic diagram for an instrumentation amplifier circuit with a low input impedance. A suitable parts list for this project appears as Table 9-7. Essentially, this circuit is just a more sophisticated version of the basic differential amplifier as discussed back in chapter 1. This circuit has been designed to emphasize the features of instrumentation amplifiers described above.

Input signal V_2 is fed into IC1A. This op amp, along with its associated components, is a simple inverting amplifier with unity gain. (The value of resistor R1 should be equal to that of resistor R2.) The inverted version of V_2 is then summed with V_1 by IC1B. This op amp (IC1B), along with its associated components, is an inverting summing amplifier. For the two differential input signals to be equally weighted, their input resistors to the summing amplifier stage (R3 and R4) should have equal values.

Feedback resistor R5 determines the gain of the differential amplifier as a whole. Using the component values suggested in

162 Miscellaneous projects

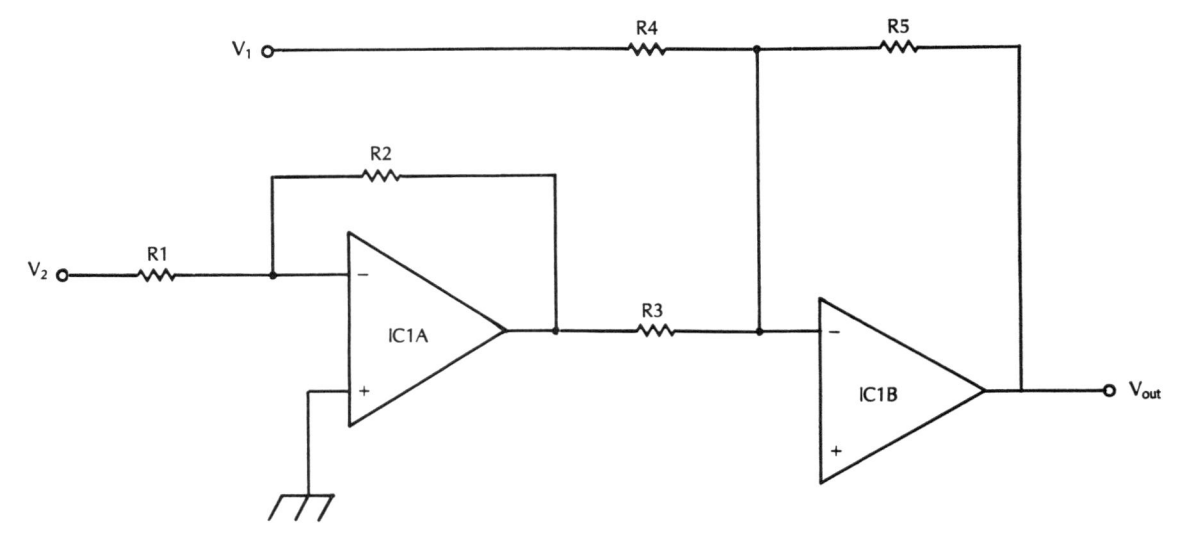

Fig. 9-14 *Project 47: low-impedance instrumentation amplifier.*

**Table 9-7. Parts List for Project 47:
Low-Impedance Instrumentation Amplifier.**

Part	Description
IC1	747 dual op amp
R1, R2, R3, R4	100K 1/4-watt 5% resistor
R5	1 Megohm 1/4-watt 5% resistor (see text)

the parts list, the circuit gain works out to:

$$G = -1{,}000{,}000/100{,}000$$
$$= -10$$

As long as resistor R3 has the same value as resistor R4, and resistor R1 equals resistor R2, both input signals will be subjected to equal gain.

In operation, input V_2 acts as an inverting input, and V_1 acts as a noninverting input. Notice that the output polarity of this circuit is inverted. If your specific intended application demands noninverted signal polarity, just add an inverting voltage follower stage to the output of this circuit.

The input impedance of this circuit is equal to the input resistor for each of the two differential inputs, resistor R1 for V_2 and resistor R4 for V_1. For maximum balance, these resistances should be equal. Since it was already determined that R1 should equal R2 and that R3 should equal R4, it turns out that all four of these resistors should have equal values. This can be stated as:

$$R_1 = R_2 = R_3 = R_4 = R$$

The only component to worry about in the circuit design is the gain determining feedback resistor R5, which can be found by rearranging the gain formula:

$$G = -R_5/R$$
$$R_5 = -G \times R$$

To prevent excessive loading, the source impedance at each input should be significantly less than the resistance of the input resistor (R).

HIGH-IMPEDANCE INSTRUMENTATION AMPLIFIER

In the vast majority of instrument amplifier applications, the highest possible input impedance is desired, for minimal loading of the source circuit. The simple instrumentation amplifier circuit shown in Fig. 9-15 features an extremely high input impedance. The input impedance of this circuit can be as high as 10 Megohms (10,000,000 ohms), or possibly more, depending on the specific op amp devices being used. A suitable parts list for this project appears as Table 9-8.

The secret of this circuit is that, unlike the low-impedance circuit of Fig. 9-14, both of the circuit's differential input are fed into the noninverting (rather than the inverting) inputs of the op amps. A noninverting amplifier circuit usually has a much higher input impedance than a comparable inverting amplifier circuit. In a standard inverting amplifier circuit, the input impedance is limited by the value of the input resistor. In a noninverting amplifier circuit the circuit's input impedance is essentially the same as the input impedance of the op amp device.

The rated input impedance for the 741 and 747 operational amplifiers is 6 Megohms (6,000,000 ohms). Bear in mind, however, that this is a typical value, rather than an absolute value. The guaranteed minimum input impedance for 741/747 type op amps is 1 Megohm (1,000,000 ohms). This is still quite high, but

164 Miscellaneous projects

Fig. 9-15 Project 48: high-impedance instrumentation amplifier.

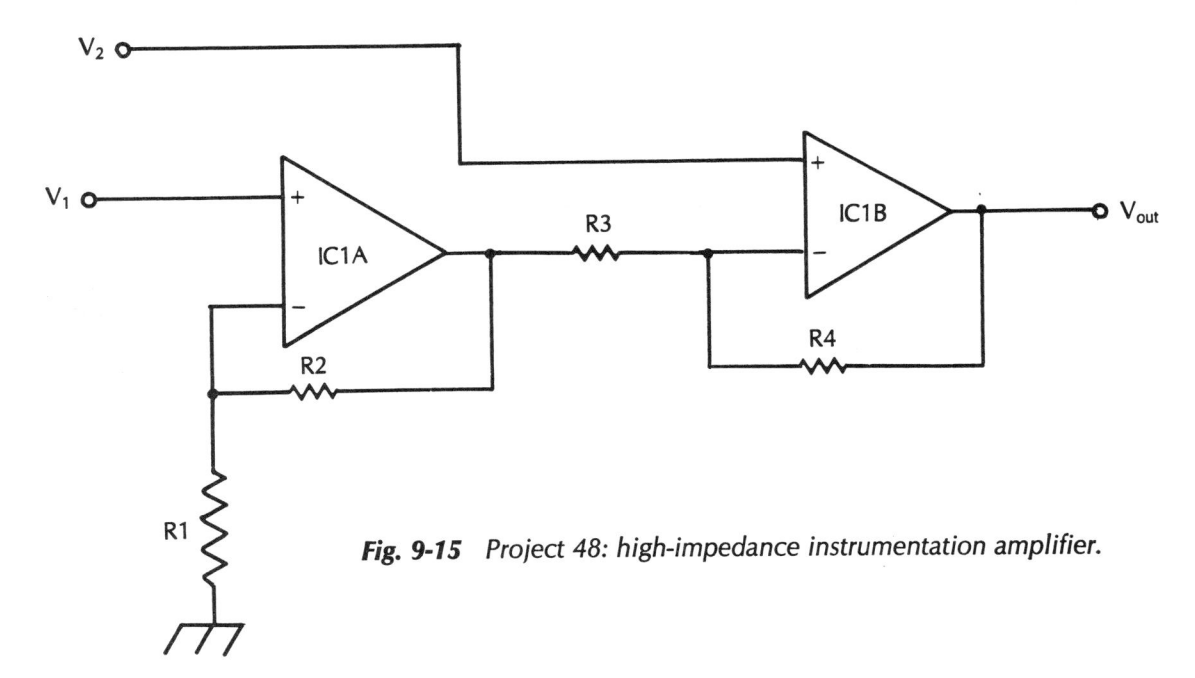

Table 9-8. Parts List for Project 48: High-Impedance Instrumentation Amplifier.

Part	Description
IC1	747 dual op amp
R1, R4	100K 1/4-watt 5% resistor (see text)
R2, R3	1K 1/4-watt 5% resistor (see text)

it might not be high enough for some critical applications. If your intended application is very critical in terms of the input impedance and loading effects, you should replace the 747 dual op amp IC called for in the parts list with higher-grade op amp devices. Operational amplifiers with FET input stages typically offer exceptionally high input impedances.

The instrumentation amplifier circuit of Fig. 9-15 is also more flexible than the instrumentation amplifier circuit of Fig. 9-14. The source impedances don't have to match precisely. A slight imbalance in the source impedances will not adversely affect the common-mode rejection of this circuit.

In this circuit, IC1A and its associated components form a simple noninverting amplifier for input V_1. IC1B and its associated components, form a differential amplifier that subtracts input V_1 from input V_2.

The gain for input V_2 is equal to:

$$G_2 = 1 + R_4/R_3$$

Using the component values suggested in the parts list, the V_2 gain works out to:

$$G_2 = 1 + 100{,}000/1000$$
$$= 1 + 100$$
$$= 101$$

The gain equation for the other input signal (V_1) is slightly more complex:

$$G_1 = -(R_4/R_3) \times (1 + R_2/R_1)$$

Again, you can easily plug in the appropriate resistor values from the recommended parts list:

$$G_1 = -(100{,}000/1000) \times (1 + 1000/100{,}000)$$
$$= -100 \times (1 + 0.01)$$
$$= -100 \times 1.01$$
$$= -101$$

Notice that this is the inverted equivalent to the gain of input V_2.

You can combine these two gain equations with the input voltages to find the resulting output voltage for this circuit:

$$V_{out} = [(1 + R_4/R_3) \times V_2] - [R_4/R_3 \times (1 + R_2/R_1) \times V_1]$$

If you assume that resistor R1 has the same value as resistor R4, and resistor R2 equals resistor R3, then the circuit is fully balanced, and the output voltage equation can be simplified to:

$$V_{out} = [(1 + R_1/R_2) \times V_2] - [R_1/R_2 \times (1 + R_2/R_1) \times V_1]$$
$$V_{out} = [(1 + R_1/R_2) \times V_2] - [(R_1/R_2 + 1) \times V_1]$$
$$V_{out} = (1 + R_1/R_2) \times (V_2 - V_1)$$

For the component values given in the parts list, this works out to:

$$V_{out} = (1 + 100{,}000/1000) \times (V_2 - V_1)$$
$$= (1 + 100) \times (V_2 - V_1)$$
$$= 101 \times (V_2 - V_1)$$

The differential voltage of the two inputs is amplified by the same amount as the individual voltages are. In applications

166 Miscellaneous projects

where this 101 gain may be critical, you can simply follow the instrumentation amplifier circuit with a suitable voltage divider network of some kind.

So what happened to the input V_2 and the input V_1. These individual signals cancel each other leaving only the difference, because the two input voltages are being combined *differentially*. Looking at the circuit as a whole, it is a differential amplifier with a total effective gain of 101—for only the differential output voltage.

This instrumentation amplifier circuit offers an extremely high input impedance, and exceptionally good common-mode rejection. However, it is limited in some applications because there is no convenient way to permit manual control over the circuit gain. This is because the resistor balances must be maintained throughout the circuit. A complex multi-unit ganged potentiometer could theoretically do the job, but that would be an awkward and highly inelegant solution at best. This leads right into the next, and final project.

ADJUSTABLE GAIN INSTRUMENTATION AMPLIFIER

The instrumentation amplifier circuit shown in Fig. 9-16 offers manual gain adjustment with a simple (single) potentiometer (R1). This comes mainly at the expense of a third operational amplifier stage (IC2). A suitable parts list for this project is given in Table 9-9.

Each of the two input signals (V_1 and V_2) are fed into separate noninverting amplifier stages (IC1A for V_1 and IC1B for V_2). The use of the noninverting inputs permits the circuit to function with a very high input impedance, as in the preceding project. The outputs of these amplifiers are then combined in a difference amplifier (IC2).

In an operational amplifier, the voltage drop between the inverting input and the noninverting input is nominally zero. This places the gain control potentiometer in a voltage divider network affecting both noninverting amplifier stages. Actually, while it might not be obvious when first looking at the schematic, this voltage divider network permits both input signals (V_1 and V_2) to pass through both of the two noninverting amplifiers in this circuit. The output voltage of IC1A is equal to:

$$V_{out1} = [(1 + R_2/R_1) \times V_1)] - (R_2/R_1 \times V_2)$$

Adjustable gain instrumentation amplifier   167

Fig. 9-16 *Project 49: adjustable gain instrumentation amplifier.*

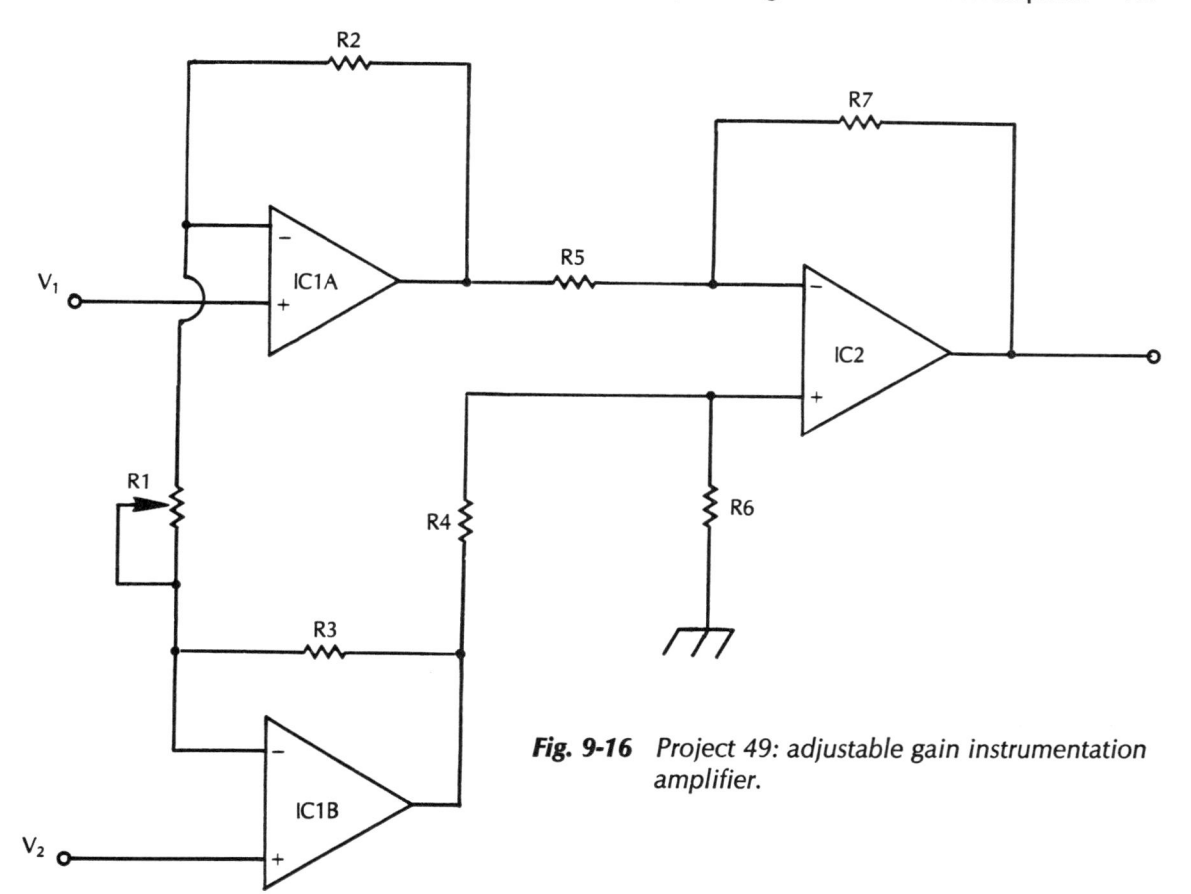

Table 9-9. Parts List for Project 49: Adjustable Gain Instrumentation Amplifier.

Part	Description
IC1	747 dual op amp
IC2	op amp ($1/2$ 747, or 741, or similar)
R1	25K potentiometer
R2, R3	47K $1/4$-watt 5% resistor (see text)
R4, R5	10K $1/4$-watt 5% resistor (see text)
R6, R7	100K $1/4$-watt 5% resistor (see text)

168 Miscellaneous projects

For the second noninverting amplifier (IC1B), the gain equation is similar, except the input signals are reversed:

$$V_{out2} = [(1 + R_3/R_1) \times V_2)] - (R_3/R_1 \times V_1)$$

Normally the two feedback resistors (R2 and R3) will have equal values. This equality will be assumed throughout the rest of our discussion of this circuit. The value of R1, of course, depends on the setting of the potentiometer.

The output signals from the two noninverting amplifier stages are fed into a difference amplifier (IC2). Again, in virtually all practical instrumentation amplifier applications, the two input resistors (R4 and R5) will have equal values. Again, this equality will be assumed in the discussion. The output of the difference amplifier is equal to its gain multiplied by the difference between its two input voltages:

$$V_{out} = R_7/R_4 \times (V_{out2} - V_{out1})$$

In combining the equations, assume that resistor R3 is the same as R2, so R2 will be used in place of R3 in the equations. This will simplify the algebra. Using the three equations given so far in this section, you can rewrite the formula for the output voltage as:

$$V_{out} = R_7/R_4 \times \{[(1 + R_2/R_1) \times V_2] - (R_2/R_1 \times V_1) \\ - [(1 + R_2/R_1) \times V_1] - (R_2/R_1 \times V_2)\}$$

This is a pretty complex-looking formula, but with a little bit of algebraic rearrangement, it can simplify to:

$$V_{out} = R_7/R_4 \times (1 + 2R_2/R_1) \times (V_2 - V_1)$$

Using the suggested component values from the recommended parts list and assuming that potentiometer R1 is set for a value of 10K (10,000 ohms), the output voltage works out to:

$$\begin{aligned} V_{out} &= 100{,}000/10{,}000 \times \left(1 + \frac{2 \times 47{,}000}{10{,}000}\right) \\ &\quad \times (V_2 - V_1) \\ &= 10 \times (1 + 94{,}000/10{,}000) \times (V_2 - V_1) \\ &= 10 \times (1 + 9.4) \times (V_2 - V_1) \\ &= 10 \times 10.4 \times (V_2 - V_1) \\ &= 20.4 \times (V_2 - V_1) \end{aligned}$$

Of course, changing the setting (and thus, the resistance value) of potentiometer R1 will alter the overall gain of the circuit.

For the best results in this circuit, it is important that the following resistor values are made as equal as possible:

$$R_2 = R_3$$
$$R_4 = R_5$$
$$R_6 = R_7$$

This is especially true for the feedback resistors in the noninverting amplifier stages (R2 and R3). If there is a minor mismatch between these resistors it will not adversely affect the common-mode rejection ratio of the circuit, however, it will throw off the gain balances, which are important in most instrumentation amplifier applications. In critical applications, it would be a good idea to use low-tolerance, precision resistors in this circuit. Alternately, you could match up the resistors with an ohmmeter for a minimum of error.

The gain of the difference amplifier stage (IC2) should be made no larger than 10 to minimize output offset errors. In many applications, it might be a good idea to set up this stage for unity gain. In this case:

$$R_4 = R_5 = R_6 = R_7$$

This versatile circuit is useful in many instrumentation amplifier applications, even when the manual gain control is not required. You could replace the potentiometer (R1) with a fixed resistor if you do not need the controllable gain function. This is still a very good instrumentation amplifier circuit.

There you have it. Almost 50 projects for the 747 dual operational amplifier IC. As you can see, while a single op amp is versatile, powerful, and useful, two (or more) are even better.

Index

A

absolute value circuit, 20-24
ac coupled bistable multivibrator, 126-128
ac voltages, 105
adjustable gain instrumentation amplifier, 166-169
amplifiers
  antilogarithmic, 39-41
  differential, 2, 9, 17
  differential inputs/outputs, 152-154
  instrumentation, adjustable gain, 166-169
  instrumentation, high-impedance, 163-166
  instrumentation, low-impedance, 161-163
  inverting, 3-5
  inverting, high input impedance, 149-151
  logarithmic, 38-39, 41
  noninverting, 6-8
  oscillation in, 59
  pre-amp, audio, 49-51
  pre-amp, stereo magnetic cartridge, 51-53
  pre-amp, tape head, 53-55
  summing, 17-20, 41
amplitude modulation (AM), 109, 110, 113
analog circuitry
  analog-to-digital (AD) converter, 132
  digital-to-analog (DA) converter, 131-135
  digital-to-analog (DA) converter, improved, 137-139
  digital-to-analog (DA) converter, R-2R, 135-137
  digital circuitry vs., 130
analog-to-digital (AD) converter, 131-135
antilogarithm, 38
antilogarithmic amplifier, 39-41
astable multivibrators, 119, 121-122, 128-129
audio frequency (AF) signal generators, 60
audio pre-amp, 49-51
audio projects, 49-58
  mixer, 55-58
  pre-amp, 49-51
  stereo magnetic cartridge pre-amp, 51-53
  tape head pre-amp, 53-55
  tone control, 54-56

B

band-reject filter, 75, 76
bandpass filter, 75, 76, 85-90
binary values, 130, 131
binary word, 131
bistable multivibrators, 68, 119
  ac coupled, 126-128
  dc coupled, 125-126
  flip-flops, 120
  memory usage of, 121
  outputs, 121
  reversing output state of, 122

Index

bits, 130
bytes, 131

C

carrier signal, modulation, 109
center frequency, filter, 86
channel separation, 16
common mode rejection ratio (CMRR), 14
comparator
  magnitude, 27-31
  window, 24-27
converter, RMS to dc, 105-108
cosine wave, 61

D

dc coupled bistable multivibrator, 125-126
dc voltages, 105
dc/RMS converter, 105-108
deadspace circuit, 154-157
decibel meter, 104-105
decimal values, 130
demodulation, 116
detector
  magnitude, 27-31
  peak, 31-34
  peak, inverting, 34-35
  peak, manual reset, 35-37
differential amplifiers, 2, 9, 17, 152-154
differential voltmeter, 99-102
digital circuitry
  analog circuitry vs., 130
  analog-to-digital (AD) converter, 132
  applications for, 131
  digital-to-analog (DA) converter, 131-135
  digital-to-analog (DA) converter, improved, 137-139
  digital-to-analog (DA) converter, R-2R, 135-137
digital-to-analog (DA) converter, 131-135
  improved, 137-139
  R-2R type, 135-137
disallowed conditions, 5-6
divider circuit, 41-45
double-ended output, 152-154
dual-polarity voltage regulator, 141-145

E

exponent circuit, 44-48

F

feedback, 59
filter, 75-97
  band-reject, 75, 76
  bandpass, 75, 76, 85-90
  center frequency of, 86
  high-pass, 75
  high-pass, fourth-order, 82-84
  low-pass, 75-77
  low-pass, first-order, 78
  low-pass, second-order, 78
  low-pass, third-order, 77-82
  notch, 76
  notch, 60 Hz, 90-92
  orders, 76, 77
  Q or quality factor of, 87
  slope, 77
  speech, 96-97
  state variable, 92-96
first-order low-pass filter, 78
flip-flops, 68, 120
fourth-order high-pass filter, 82-84
frequency modulation (FM), 109, 110
  signal generator for, 110-113
frequency response, 16
full-wave rectifier, precision, 141-149
function generators, 66
  state variable filter vs., 92
fundamental frequency, 59
gain
  open loop, 741 op amps, 13-14
  open loop, op amps, 3, 10

H

harmonics, 59
high input impedance inverting amplifier, 149-151
high-impedance instrumentation amplifier, 163-166
high-pass filter, 75
  fourth-order, 82-84
hiss, 16

I

impedance, op amps, 10
input impedance, 10
instrumentation amplifier
  adjustable gain, 166-169
  high-impedance, 163-166

low-impedance, 161-163
inverting amplifier, 3-5
  high input impedance, 149-151
inverting input, op amps, 2-3, 17
inverting peak detector, 34-35
inverting voltage follower, 5-6, 41

L

LED null indicator, 103-104
limiter, series, 157-161
linear circuitry (*see* analog circuitry)
logarithmic amplifier, 38-39, 41
logarithms, 37-40
low-impedance instrumentation amplifier, 161-163
low-pass filter, 75-77
  first-order, 78
  second-order, 78
  third-order, 77-82

M

magnitude detector, 27-31
mathematical operations circuits, 17-48
  absolute value circuit, 20-24
  anitlogarithmic amplifier, 39-40
  divider circuit, 41-45
  exponent circuit, 44-48
  inverting peak detector, 34-35
  logarithmic amplifier, 38-39
  magnitude detector, 27-31
  multiplication circuit, 40-43
  peak detector, 31-34
  peak detector, manual reset, 35-37
  summing amplifier, 17-20
  window comparator, 24-27
memory, bistable multivibrators as, 121
meters, decibel, 104-105
mixer, 55-58
modulating signal (*see* program signal)
modulation
  amplitude (AM), 109, 110, 113
  carrier signal in, 109
  demodulation and, 116
  frequency (FM), 109, 110
  program signal in, 109
  pulse amplitude (PAM), 110, 113-117
  pulse amplitude (PAM), demodulator circuit, 116-117
  pulse width (PWM), 110, 117-118

modulators, 109-118
  FM signal generator, 110-113
  pulse amplitude demodulator, 116-117
  pulse amplitude modulator, 113-117
  pulse width modulator, 117-118
monostable multivibrators, 119
  one-shot, 119, 123-125
  pulse stretchers, 120
  timers, 120
multiplication circuit, 40-43
multivibrator, bistable, 68
multivibrators, 119
  ac coupled, 126-128
  astable (*see* astable multivibrators)
  bistable (*see* bistable multivibrators)
  dc coupled, 125-126
  monostable (*see* monostable multivibrators)
  one-shot timer, 123-125
  trigger pulse, 119

N

nibbles, 131
noisy operation, 16
noninverting amplifier, 6-8
noninverting input, op amps, 2-3, 17
noninverting voltage follower, 8-9
notch filter, 76
  60 Hz, 90-92
null indicator, LED, 103-104
null voltmeter, 102-103

O

one-shot timer (multivibrator), 119, 123-125
op amps, 1
  741-type, 10-14
  747, 14-16
  applications for, 1
  channel separation, dual packages, 16
  common mode rejection ratio (CMRR), 14
  development of, 1
  differential amplifier function of, 2, 9, 17
  disallowed conditions and, 5-6
  dual-packaged, 1-2
  function of, 2-10
  input impedance, 10

Index

op amps (con't.)
  inverting amplifier using, 3-5
  inverting voltage follower, 5-6
  noninverting amplifier using, 6-8
  noninverting voltage follower using, 8-9
  open loop gain, 3, 10
  output impedance, 10
  output signal of, 2
  power supplies, dual, 10
  schematic symbol for, 2
  slew rate, 14
  voltage inputs, inverting and noninverting, 2-3, 17
open loop gain, 10
  741 op amps, 13-14
  op amps, 3
orders, filter, 76, 77
oscillators, 59
  quadrature, 60-63
  state variable filter vs., 92
  two-frequency, 66-69
output signal, op amps, 2
output, double-ended, 152-154

P

peak detector, 31-34
  inverting, 34-35
  manual reset, 35-37
peak voltage, 106
peak-to-peak voltage, 106
power supplies
  741 op amps, 13-14
  747 dual op amps, 14-15
power supplies, op amps, 10
pre-amp
  audio, 49-51
  stereo magnetic cartridge, 51-53
  tape head, 53-55
precision full-wave rectifier, 141-149
program signal, modulation, 109
pulse amplitude modulation (PAM), 110, 133-117
  demodulator circuit, 116-117
pulse circuits, 119-139
  ac coupled bistable multivibrator, 126-128
  analog-to-digital (AD) converter, 132
  astable multivibrator, 128-129
  dc coupled bistable multivibrator, 125-126
  digital vs. analog circuitry, 130

digital-to-analog (DA) converter, 131-135
digital-to-analog (DA) converter, improved, 137-139
digital-to-analog (DA) converter, R-2R type, 135-137
improved DA converter, 137-139
one-shot timer, 123-125
R-2R DA converter, 135-137
slew rate, 119
transition time between pulses, 119
pulse stretchers, monostable multivibrators, 120
pulse width modulation (PWM), 110, 117-118

Q

Q (quality) factor, filter, 87
quadrature oscillator, 60-63

R

radio frequency (RF) signal generators, 60
ramp wave (see sawtooth wave)
random noise, 16
rectifier, full-wave, precision, 141-149
RMS to dc converter, 105-108
root-mean-square (RMS) voltage, 106

S

60 Hz notch filter, 90-92
741 op amps, 10-14
  8-pin round can package, 12
  common mode rejection ratio (CMRR), 14
  dual inline package (DIP), 11, 13
  frequency compensation, internal, 12
  frequency response in, 16
  noisy operation, 16
  open loop gain for, 13-14
  pinouts for, 11-13
  power supply for, 13-14
  schematic symbol for, 11
  slew rate, 14
747 dual op amps, 1, 14-16
  channel separation, 16
  frequency response in, 16
  noisy operation, 16
  pinouts for, 15
  power supply for, 14-15
sawtooth wave generator, 69-73

Index *175*

sawtooth waves, ascending vs. descending, 69-73
schematic symbols
  741 op amps, 11
  op amps, 2
second-order low-pass filter, 78
series limiter, 157-161
signal generators, 59-73
  amplifiers and, 59
  audio frequency (AF) vs. radio frequency (RF), 60
  frequency modulation (FM), 110-113
  frequency range of, 59
  function generators, 66, 92
  fundamental frequency and harmonics, 59
  oscillations in, 59
  quadrature oscillator, 60-63
  sawtooth wave, 69-73
  square wave, 64
  sub-audio, 60
  triangle wave, 63-66
  two-frequency oscillator, 66-69
sine wave, 61
slew rate
  741 op amps, 14
  pulse circuits, 119
slope, filter, 77
speech filter, 96-97
square wave generator, 64
state variable filter, 92-96
stereo magnetic cartridge pre-amp, 51-53
sub-audio signal generators, 60
summing amplifier, 17-20, 41

T

tape head pre-amp, 53-55
test equipment, 99-108
  decibel meter, 104-105
  differential voltmeter, 99-102
  null indicator, LED, 103-104
  null voltmeter, 102-103
  RMS to dc converter, 105-108
thermocouples, 106-108
third-order low-pass filter, 77-82
timers, monostable multivibrators, 120
tone controls, 54-56
triangle wave generator, 63-66
trigger pulse, 119
two-frequency oscillator, 66-69

V

voltage follower
  inverting, 5-6, 41
  noninverting, 8-9
voltage regulator, dual-polarity, 141-145
voltmeters
  differential, 99-102
  null, 102-103

W

window comparator, 24-27
word, binary, 131

Other Bestsellers of Related Interest

BUILD YOUR OWN LASER, PHASER, ION RAY GUN & OTHER WORKING SPACE-AGE PROJECTS
—Robert E. Iannini

Here's the highly-skilled do-it-yourself guidance that makes it possible for you to build such interesting and useful projects as a burning laser, a high power ruby YAG, a high-frequency translator, a light beam communications system, a snooper phone listening device, and more—24 exciting projects in all! 400 pages, 302 illustrations. **Book No. 1604, $17.95 paperback, $24.95 hardcover**

BUILD YOUR OWN WORKING FIBEROPTIC, INFRARED AND LASER SPACE-AGE PROJECTS
—Robert E. Iannini

Here are plans for a variety of useful electronic and scientific devices, including a high sensitivity laser light detector and a high voltage laboratory generator (useful in all sorts of laser, plasma ion, and particle applications as well as for lighting displays and special effects). And that's just the beginning of the exciting space age technology that you'll be able to put to work! 288 pages, 198 illustrations. **Book No. 2724, $16.95 paperback, $24.95 hardcover**

BASIC ELECTRONICS THEORY—3rd Edition
—Delton T. Horn

"All the information needed for a basic understanding of almost any electronic device or circuit . . ." was how *Radio-Electronics* magazine described the previous edition of this now-classic sourcebook. This completely updated and expanded edition provides a resource tool that belongs in a prominent place on every electronics bookshelf. Packed with illustrations, schematics, projects, and experiments, it's a book you won't want to miss! 544 pages, 650 illustrations. **Book No. 3195, $21.95 paperback, $28.95 hardcover**

THE THYRISTOR BOOK—with 49 Projects
—Delton T. Horn

With this collection of 49 projects, Delton T. Horn effectively demystifies the thyristor, explaining in simple terms the theory of thyristor construction and operation. You will learn the secrets of silicon-controlled rectifiers, triacs, diacs, and quadracs. You will also get dozens of practical examples for their use, including: light dimmer, self-activating night light, timed switch, visible doorbell, touch switch, and electronic crowbar. 220 pages, 153 illustrations. **Book No. 3307, $16.95 paperback, $26.95 hardcover**

THE COMPARATOR BOOK—with Forty-Nine Projects—Delton T. Horn

Horn sparks new interest in this low-cost device and offers hands-on applications as well as useful in-depth theoretical background. Step-by-step instructions, detailed diagrams, and complete parts lists are provided for each project. You don't have to be an expert, either. The projects gradually become more complex; those presented later in the book build on skills you developed while working on earlier ones. 200 pages, 155 illustrations. **Book No. 3312, $16.95 paperback, $23.95 hardcover**

GORDON McCOMB'S GADGETEER'S GOLDMINE!: 55 Space-Age Projects
—Gordon McComb

This is one of the most exciting collections of electronic projects available anywhere, featuring experiments in everything from magnetic levitation and lasers to high-tech surveillance and digital communications. Find instructions for building such useful items as a fiberoptic communications link, a Geiger counter, a laser alarm system, and more. All designs have been thoroughly tested. Suggested alternative approaches, parts lists, sources, and components are also provided. 432 pages 274 illustrations. **Book No. 3360, $18.95 paperback, $29.95 hardcover**

ENCYCLOPEDIA OF ELECTRONICS—2nd Edition
—*Stan Gibilisco and Neil Sclater, Co-Editors-in-Chief*
Praise for the first edition:
". . . a fine on-volume source of detailed information for the whole breadth of electronics."
—*Modern Electronics*

The second edition, newly revised and expanded, brings you more than 950 pages of listings that cover virtually every electronics concept and component imaginable. From basic electronics terms to state-of-the-art applications, this is the most complete and comprehensive reference available for anyone involved in any area of electronics practice! 976 pages, 1400 illustrations. **Book No. 3389, $69.50 hardcover only**

BASIC DIGITAL ELECTRONICS—2nd Edition
—*Ray Ryan and Lisa A. Doyle*

An easy-to-follow digital electronics reference guide. This has long been the key to analyzing and understanding digital circuitry. Now this popular reference includes microprocessor basics, alphanumeric codes, conversions, and number systems; explanations of the basics of logic gates, families, and networks; and example applications. 256 pages, 166 illustrations. **Book No. 3370, $16.95 paperback, $25.95 hardcover**

Prices Subject to Change Without Notice.

Look for These and Other TAB Books at Your Local Bookstore

To Order Call Toll Free 1-800-822-8158
(in PA, AK, and Canada call 717-794-2191)

or write to TAB BOOKS, Blue Ridge Summit, PA 17294-0840.

Title _____ Product No. _____ Quantity _____ Price _____

_____

_____

_____

☐ Check or money order made payable to TAB BOOKS

Charge my ☐ VISA ☐ MasterCard ☐ American Express

Acct. No. _____ Exp. _____

Signature: _____

Name: _____

Address: _____

City: _____

State: _____ Zip: _____

Subtotal $ _____

Postage and Handling
($3.00 in U.S., $5.00 outside U.S.) $ _____

Add applicable state and local sales tax $ _____

TOTAL $ _____

TAB BOOKS catalog free with purchase; otherwise send $1.00 in check or money order and receive $1.00 credit on your next purchase.

Orders outside U.S. must pay with international money order in U.S. dollars.

TAB Guarantee: If for any reason you are not satisfied with the book(s) you order, simply return it (them) within 15 days and receive a full refund.
BC